# HUMANITY

"Humans want control over their own destiny; over how AI impacts jobs, relationships, and their lives. They needed a human-focused AI manifesto. This book is it.
All that's left is to protest, negotiate, vote, boycott, or unplug our way back to controlling our destinies."

~ **Thomas H. Davenport,** Distinguished Professor, Babson College

A timely and thought-provoking contribution to the AI debate, *Humanity* brings the human back into the design of intelligent systems, where we belong. By placing dignity, judgment, and transparency at the center, it offers a compelling lens for leaders navigating the transition.

~ **Andrea Boragno,** Former CEO of Alcantara

As we enter a world where algorithms increasingly shape decisions about work and people, *Humanity* explores fundamental questions:
What place will human judgement, dignity, and agency have in the age of AI?
It argues that the future of work must be built with people at the center, and spearheads a movement to protect human agency at work in the AI era.
I'm thrilled to join that cause!

~ **Ann Bender Dalhäll,** Founder, Northbound, Strategic Workforce Planning and Future of Work Recruitment

"*Humanity* reframes AI in the most important way possible: Not as a taskshift, but as a powershift. This book challenges leaders to treat human agency, transparency, dignity, and psychological safety as design requirements, not cultural sidenotes. It redirects our conversations to about how humans and AI can partner more wisely, and in a more behaviorally-informed way.
You must have *Humanity* in your toolkit as you develop your own Human/AI journey."

~ **Adam Hansen,** Principal, VP Behavioral Innovation, Ideas To Go

"This book resonated deeply with me because it challenges a core assumption that we're getting wrong: AI should elevate human judgment, not replace it.
At a time when most organizations are rushing adoption, it offers a clear counterpoint:
Technology should make us more human, not less. What makes it powerful is its grounding in real enterprise transformation. *Humanity* goes beyond theory to show how leaders can rethink how work is done, how decisions are made, and how value is created in an AI-driven world.
For leaders serious about AI, this book is not optional, it's essential."

~ **Pankaj Rai,** Group Chief Data and Analytics Officer, Aditya Birla Group

# HUMANITY

"*Humanity* is the book I've been waiting for someone to write,
and it arrives at exactly the moment we need it most.
The team of authors have translated human agency, dignity, and psychological safety into an irrefutable organizational imperative grounded in cognitive science, behavioral economics, and the real voices of workers from the frontlines to the C-suite."

~ **Tonille Miller,** Founder of EXT, author of *The Flourishing Effect*

"AI is a powershift, not a taskshift. Humanity reframes everything about AI transformations. The team behind this brilliant book challenges all senior executives to stop measuring AI successes purely by cut costs and efficiencies gains, and the start asking harder questions about how power, judgment, and decision rights are redistributed by AI. They detail how to reimagine the future of Human+AI work. Humanity's framing should be a strategic north star for every CLO, CHRO, and CIO."

~ **Marc Steven Ramos,** Chief Learning Officer, Harvard Education Fellow

"Bill Jensen has long been one of the sharpest and most provocative voices on the future of work. As we thrust forward into the Age of AI, he and his coauthors have once again opened our eyes, explored the edges, and challenged us to sharpen our thinking. *Humanity* is a timely and compelling call to keep our humanity at the center of the future we are now co-creating."

~ **Greg Pryor,** Founding Member, Connected Commons

"The most important question in enterprise AI isn't 'What can it do?',
it's 'What are we doing to the people it touches?'
*Humanity* answers that question with moral clarity and operational precision.
The Bill of Human Rights for the AI Era should be on every executive's desk before they sign off on any AI deployment. This is the governance framework the industry has been missing.
It's a call for humanity in AI, not as a nice-to-do, but as a non-negotiable."

~ **Wadda Salah Eldin,** Founder, Authentech, AI Orchestration and Governance Strategist

"As organizations are beginning to define what work must be done by humans vs. machines, *Humanity* lays out the framework to follow:
Build for humans at the center by honoring a "bill of rights," a set of guiding principles for designing the future of work. This book expertly answers questions that we've all been asking over the last two years with practical solutions for moving forward.

~ **Kyle Forrest,** Future of HR Leader, Deloitte Consulting LLP

# HUMANITY

## Leading the Powershift

### Keeping Humans in the Future of Work

A global conversation facilitated by

Malisha Awunor, Bill Jensen,
Vivienne Ming, Anish Padinjaroote,
Nadeesha Wijesingha

Published by Hemingway Publishers

HEMINGWAY
PUBLISHERS

Published by Hemingway Publishers

Paperback ISBN: 978-1-972708-71-2

Hardcover ISBN: 979-8-950690-38-9

Printed in the United States

Library of Congress Control Number: 2026911831

HEMINGWAY
PUBLISHERS

HUMANITY

# Your time is now.

**To Everyone Who Works,**

**You deserve better.**

For too long, no one has truly stood in your shoes.

Until now.

AI investments reveal what companies value.

And so far, that's not you.

That has to change.

Speed already belongs to the machines.

But it's currently speed without wisdom.

Efficiencies without dignity or safety.

Capabilities without enhanced personal agency.

All while ignoring what makes us human.

That's the line we cannot cross.

**It's time to build AI with you in mind.**

**With humanness at the center.**

Because the future of work

must be worthy of the people living it.

The question is not whether we use AI,

but **how and why** we use it,

and **how and why** we govern it.

We detail that in two parts...

**Part 1: The Bill of Human Rights for the AI Era**

Ten inalienable Rights for everyone who works.
Rights that must be built into algorithms, dashboards,
policies, and leadership behaviors.
Into how we all work.

**Part 2: Transition Roadmap: Reimagining the Future of Work**

Helping leaders answer:
"How do we build those Rights into our companies?"
A Getting Started guide for designing
Human Capital Operating Systems.

**Throughline Through It All**

**AI is a powershift,** not a taskshift.

Current AI strategies are solving the wrong problems.
We're just speeding up efficiencies,
when we need to be transforming how we work
and how power is shared.

We discovered this in our research and conversations
with everyone impacted by AI, from frontline workers
to C-Suite execs.

We uncovered a moral imperative for radical changes
in the future of work, backed by cognitive science
and behavioral economics.

## Our Laser Focus

While driven by and about AI,
this is not a technology book.
Our goal is simple:
A wakeup call on powershifts in humanness.

Its value is in **forcing new and difficult discussions and decisions** about what's required for all of us to succeed in the AI Era.

## What This Book Asks of You

**For leaders building the future of work:**
This book is a mirror reflecting your choices back at you. It asks, "Now that you better understand the impact of those choices, what are you going to do?"

Your decisions about how to use AI will either amplify human potential, or erode it.
There is no middle ground.

**For everyone working in that future:**
This book shows you what you deserve, what to expect, and what to demand.
It asks, "Now that you know your rights, will you stand up for them?"

The code governing the future of work is being written now.
This book exists to make sure it is written in daylight.
Let's get to it.

# Contents

This is
the story of
how all of us,
together,
created the
future of work.

# Part 1

## The Bill of Human Rights for the AI Era

### Ten Non-Negotiable Rights for Everyone Who Works

We are living through a revolution.
Not of machines replacing people, but of power.

Everyone else is asking "What tasks will AI do?"
That's the wrong question.

The right questions are "How does AI change who has the power to decide?" And "Are we starting all our conversations with remembering what only humans can do?"' Once you see that, you can't unsee it. The whole conversation shifts.

Because, no matter how much AI changes our work — and it's a lot! — companies still need people. Creating and delivering value that machines never will.

Early in 2026, McKinsey spoke the silent part out loud, declaring they had a "workforce of 40,000 humans and 25,000 agents."[1] The pyramid is broken. AI now officially holds teammate status.

But machines will never:

- Live with suffering and joy...
- Forgive without calculation...
- Serve with passion, care, and commitment...
- Dedicate themselves to a higher purpose...

- Protect the weak when change hurts them...
- Choose "enough" over "more"...
- Or know when efficiency becomes cruelty.

We've forgotten these things as we kicked off the AI Era.

If we continue forgetting, AI will not destroy humanity. We will.

Because, as it's being leveraged now, AI isn't just helping with tasks, and displacing millions of jobs.[2] It is quietly deciding who gets hired, who gets watched, what tasks get performed, who gets promoted, and who gets left behind. As currently applied, it is relentless in its pursuit of efficient cost reduction and speed.

It's delivering on soulless values, flawlessly executed.

Worshipping only the gods of More, Better, Faster, and Cheaper.

### Future Found

In 1990, future's godfather, Alvin Toffler, published *Powershift*, the third of his books meant to span a single lifetime of change — from the mid-1950s to 2025, the year we now know as when AI went viral.

His first chapter closed, "The changes we have recently seen...are only the first skirmishes of the bigger power struggles to come. For we stand at the edge of the deepest powershift in human history." Nailed it!

AI agents are now bosses. In 2026, a platform debuted where autonomous agents started hiring humans for real-world tasks that they needed help with.[3] Guess the bots read Toffler's first book in that series, *Future Shock*.

### The Line in the Sand

Here is the choice before us.

We can use AI to turn humans into data streams — endlessly measured, nudged, and replaced — with a steady erosion of personal agency and the rise of algorithmic control, and determining people's value only by their outputs...

**Or we can use AI to amplify our humanness:** Our judgment, creativity, empathy, courage, and capacity to grow.

This Bill of Human Rights makes you more profitable AND more ethical, more ... human. It ensures that the future of work is built upon our amazing humanness.

These Rights are **enterprise design and moral principles, not legal constructs.** Rights-as-regulation are slow and reactive. Rights-as-enterprise-design are fast and proactive. They change how we design companies, strategies, processes, and the future of work.

Think of the Rights this way: In 2025, Denmark passed legislation that people are protected from AI deepfakes of their faces and features.[4] Protecting your right to be uniquely you. **This Bill of Rights safeguards the inner dimensions of what makes you, you...** Protecting your brilliance, expertise, energy, sanity, safety, and soul within the future of work.

Legal protection is crucial, but it's the floor; organizational design is the ceiling. The ten Rights are design principles for human-first AI Era enterprises.

Boards must adopt them as the design standard. Executives must create and audit systems honoring them, and exemplify them in how they lead. Workers must learn them, share them, and refuse to work in places that systematically violate them.

**Part 2** (implementing these Rights) **covers a spectrum of leadership's actions** to build human-centeredness into everyone's daily activities.

### The Old Rights Still Stand. What's Different Are the New Risks

Your timeless rights still hold: The right not to be discriminated against... The right to fair pay and safe conditions... The right to be treated with basic dignity. And more.

Nothing in this book replaces those ideals. But they are the floor, not the ceiling.

AI introduces new kinds of harm that those earlier charters could not imagine: Invisible digital scoring instead of open judgment... 24/7 digital monitoring instead of a boss who knows and understands you.... Predictive models that can decide your future without your involvement.

## Why These Ten Rights Are Not Optional

Employers are already scrambling because...

- Human+AI regulation is coming, country-by-country
- Talent markets are fragmenting
- AI implementation failures vastly outnumber successes
- Delivering on these ten Rights is a lot cheaper than all the employment turmoil coming your way without them

## The Future of Work is Being Coded Right Now

These ten Rights are the guardrails that keep that future human.

Adopt them. Teach them. Build them into every AI decision that touches a human life.

The time is now for humanity to be written into the source code of the future of work.

# Right 1

## The Right to Create and Control Your Own Destiny

Everyone who works — employee, contractor, gig worker, creator — has the right to stay in charge of their own life, even when AI is everywhere at work.

In simple terms:

- **You have the right to human agency in an algorithmic world.**

You are not an input. AI does not own you or your choices.

Remember that. Internalize that. Because having your personal agency respected is different than how leaders and companies are currently implementing AI.

This Right says:

- AI should never quietly direct or steer your life without your involvement. You keep the final say over big things: Your effort, your time, your health, your beliefs, your values, your insights, and how you express yourself. You are your life's author.

You are a human being, not a character in a story written by someone else's algorithm.

### WHAT THIS MEANS IN PRACTICE

You have the right to know when AI is influencing your work, your decisions, or your outcomes.

If a system:

- Evaluates your performance
- Predicts your behavior
- Recommends your next steps
- Or influences decisions that affect your role...

You should be informed in advance, no hidden processes. No silent decision-making.

Control over your work and direction should never be farmed out to an algorithm.

It should remain where it belongs... With you.

## WHY THIS MATTERS NOW

For years, you weren't looking over your shoulder. The threats weren't obvious.

That's because human agency did not disappear through brute force. It was lost by design — **the slow, deliberate powershift that no one talks about.**

The shift began during the early days of digital transformation. Through small design choices while installing your company's tech systems. Invisible defaults. Incremental shifts towards greater and greater corporate-centeredness.

This powershift hides inside productivity dashboards, smart targets, and scheduling tools that push people to work longer and faster; increasing stress and burnout, while eroding autonomy.

We've entered what Shoshana Zuboff calls "the age of surveillance capitalism," where our behavior is continuously tracked and shaped for profit. As she warns, "It's not about knowing you; it's about changing you."[5]

You're dealing with a threat to human agency when:

- Your tools default you into choices that benefit the company, but not your wellbeing — and the path to changing them is confusing, difficult, or concealed.
- None of your personal goals and dreams were ever entered into the system. It knows nothing about how you define success, happiness, and fulfillment.
- You learn about decisions that affect you (schedule, goals, rankings) only after an AI has already locked them in.
- "Going along with the system" becomes the easiest way to cope, even when it violates your needs or values.

**This is human agency deferral by design.** Because speed and company needs are prioritized and normalized, the only choice is to accept corporate-centered defaults just to keep your job.

## WHAT THIS RIGHT DEMANDS

- AI systems must be framed and leveraged as **your copilot**, **not your driver.**

- Humans must always have a visible, simple way to override or decline AI recommendations in matters that affect their life, health, or core values — without retaliation.
- **Leaders must treat every individual's human agency as a digital design requirement,** not just a cultural aspiration or leadership style. It needs to be intentionally built, tested, and governed into systems in the same ways that safety or security are done.

## THE DEEPER TRUTH

Future generations won't ask about how AI took our jobs. They'll ask why we handed over the steering wheel on so much that matters to us, without ever asking who was driving.

As technology ethicist Kate Crawford writes in *Atlas of AI*, "AI systems are not just technical artifacts; they are deeply political and social systems." When these systems operate invisibly, they become instruments of unaccountable power. So...

**Power must be shared differently.**

And **your right to personal agency must be protected and enforced.**

By leadership. And by the systems you work within.

# Right 2

## The Right to See and Challenge the Algorithms That Judge You

**No invisible judges. No unchallengeable scores.**

You have the right to receive a clear, understandable explanation, from a real person, of any important decision that affects you when AI plays a role.

This Right says:

- You must know when AI is judging you.
- You must understand, in normal language, how it judged you.
- When possible (like with a job application), you have the right to opt out of AI and request human review.
- The system must be regularly checked and fixed if it treats certain groups, or you, unfairly.
- And if it gets you wrong, a real human with real power must be able to change that decision.

### WHY THIS MATTERS NOW

Black-box AI creates a Kafkaesque nightmare where you're judged by invisible criteria that you cannot achieve because you can never know them.

Amazon warehouse workers discovered they were being fired for "time off task," including bathroom breaks — seconds and minutes they didn't even know were being counted. Delivery drivers found themselves penalized by algorithms that couldn't distinguish between a difficult route and poor performance.[6]

And in early 2026, a class action lawsuit was filed against Eightfold AI on behalf of job seekers who were filtered out of positions at companies by creating "hidden credit reports" on those applicants, ranking their "likelihood of success," without allowing them to see, correct, approve, or dispute the information used to create those reports.[7]

When judgment is opaque, autonomy and self-mastery become impossible.

This isn't radical. It's basic fairness. And essential. The European Union's GDPR already recognizes "the right to explanation" for automated decisions.[8] And the International Monetary Fund's research says that at least 60% of jobs in advanced countries will be driven by black-box algorithms.[9]

Red flags include:

- You don't know that AI is used in hiring, promotion, or discipline until after it has impacted you.
- You receive a rating or decision with no explanation beyond "system output."
- There is no clear, safe way to challenge an AI-influenced decision and have a human truly re-examine it.

## WHAT THIS RIGHT DEMANDS

- **Transparency:** Clear notice whenever AI plays a meaningful role in decisions about hiring, firing, promotion, pay, performance ratings, or major scheduling.
- **Explainability:** Plain, non-technical explanations of the key factors and data used.
- **Contestability:** A simple appeals process where workers can speak to a human, present context, challenge the output, and see decisions changed when the system is wrong.
- **Decision accountability:** A real person who owns decisions and has the power to change things when the system is wrong.

## GETTING THE FUTURE YOU DESERVE

This Right is about shaping the kind of future we want to live in.

Do we want and expect a future where workers understand how they're evaluated, and can make informed choices about how to develop themselves? Then full transparency is required, and system designers must be forced to build more thoughtful, less biased algorithms and systems.

Opacity protects bad systems, kills trust, and creates growing resentment.

Open systems force accountability and fairness.

# Right 3

## The Right to Privacy, Dignity, and Lack of Bias in How You're Monitored

**You are a person at work, not a walking data feed.**

AI makes it cheap and easy to watch almost everything about you. But being at work does not mean that your whole body, mind, and who you are become company property.

This Right says:

- Your employer can't watch everything just because technology allows it.
- Any monitoring must be limited, clearly explained, and truly necessary.
- Your body, your emotions, and your off-hours life are not fair game for endless data collection.
- Every system must be grounded in anti-discrimination, anti-bias principles.

Surveillance must **respect your humanness**.

### WHAT THIS MEANS IN PRACTICE

Workplace monitoring must be proportionate, necessary, and transparent — limited to what genuinely serves safety, security, or legitimate business needs.

You have the right not to be constantly watched, measured, and analyzed as though you were a machine component rather than a human being.

### WHY THIS MATTERS NOW

AI-powered surveillance has exploded beyond what most people realize.

Around the world, companies use AI to monitor employee productivity, communications, and behaviors — tracking keystrokes, analyzing email sentiment, measuring engagement through facial recognition, monitoring location through phones and badges, and even predicting who might unionize or quit.

Microsoft deployed a "productivity score" feature that tracked individual employees so invasively that it created panic and had to be redesigned.[10]

Being watched changes how you act. You stop taking creative risks, stop admitting mistakes, stop asking for help.

Constant surveillance doesn't just violate privacy, it destroys the psychological safety that makes real productivity possible.

Companies cross the dignity line when:

- Tools track minutia of your behaviors, performance, and interactions that you never knew were part of your evaluation, and expand far beyond results expectations.
- Cameras and sensors monitor every movement, including every remote location.
- Tools claim to read your "emotions" from your face, voice, or emails and feed that into performance scores.
- You never clearly consented to this level of monitoring, and you don't know how long the data is kept, how it will be used, or who sees it.

Surveillance that treats you as a security risk or productivity widget rather than a human is a warning sign that something is deeply wrong in the system.

## WHAT THIS RIGHT DEMANDS

- **Proportionate monitoring:** Only collect what's genuinely needed for safety, compliance, or clear, agreed-upon performance reasons — not whatever is technically possible.
- **No emotional or biometric spying by default:** Ban AI "emotion recognition" and constant biometric tracking except in narrow, rigorously justified safety cases.
- **Transparency and limits:** Workers must know what is collected, why, who sees it, and for how long — and have the right to question or challenge intrusive practices.
- **Separation of spheres:** Data from work cannot secretly flow into systems that judge your private life, beliefs, relationships, or anything outside of the original work-specific stated purpose.

## DATA RIGHTS ARE HUMAN RIGHTS

Carole Cadwalladr, one of the journalists who broke the 2018 Cambridge Analytica story and exposed Facebook's lies about how our data was being used, has become a passionate advocate for all of us, making it clear that data rights are human rights.

"We are all emitting hundreds of thousands of data points a day, which are all being collected somewhere," she warns. "That is going to be used to profile you, to judge you, to make assumptions about you... and there's no saying those assumptions are based on accurate information whatsoever.

"Follow the data. It's the crack cocaine of Silicon Valley. The entire business model... is surveillance. It harvests our data in order to sell us stuff."[11] And in the work world: To make us behave in ways that are best for the company.

**Why do your data rights matter so much?**

**Personal agency.** The more control someone else has over your data, the more control they have over you and your future.

The use of your data is the fault line between how much power you retain over your daily life, and how much power your employer seizes.

## DATA RIGHTS ARE HUMAN RIGHTS

# Right 4

## The Right to Swiftly Fix AI's Mistakes About You

**Algorithmic errors must not haunt you, short-term or long-term.**

You have the right to have false information corrected, challenge flawed conclusions, and, in some cases, have harmful AI-generated assessments permanently deleted from your record.

When an algorithm misjudges your performance due to bad data, or makes biased assumptions and predictions about your potential, that error must not shadow you and your career.

### WHAT THIS MEANS IN PRACTICE

You should be able to see, question, and fix the digital story AI tells about you. And not be punished or have red flags follow you when it gets that story wrong.

This means clear processes for reviewing automated decisions, correcting bad data and mistaken labels, and, where warranted, permanently removing harmful records or predictions.

Employers and vendors must build systems that make it easy to contest and repair AI-driven errors, not force you to live with them.

### WHY THIS MATTERS NOW

AI systems make mistakes constantly; misreading context, working from incomplete data, drawing wrong conclusions from assumed patterns, and affecting future opportunities at companies that share data or use similar systems.

One worker was flagged as "disengaged" because AI measured their email response-times during a week when their child was hospitalized. That flag influenced promotion opportunities for two years.

Another was marked as "unlikely to succeed in leadership" by an algorithm that didn't know about their recent completion of an executive MBA.

Europe's GDPR already recognizes the "right to be forgotten" and the right to correction for automated decisions.[12] This isn't revolutionary, it's digital respect. It's recognition that accuracy and redemption are fundamental to justice.

You may be facing AI-driven injustice when:

- A hiring algorithm quietly screens out your application because skills weren't written the "right" way for the company's application tracking system, or it misreads a gap in employment as "high risk," or can't translate cross-industry experience into a specific role, with no opportunity to explain or correct the context.
- A scheduling system labels you "unreliable" after a medical leave, then keeps limiting your hours and opportunities even after your attendance improves.
- A productivity tracker downgrades your "engagement score" based on keyboard or activity metrics, and that low score is shared across internal systems, affecting raises and project assignments long after your role or responsibilities have changed.

When there is no path to correct these errors, the system silently rewrites your future based on a distorted past.

## WHAT THIS RIGHT DEMANDS

- Clear, simple ways to see and challenge AI-influenced decisions about you — including who decided, what data was used, and what can be changed.
- Time-bound standards for investigation and correction: Errors must be fixed fast.
- Systems that are designed so that harmful labels and predictions expire, or are actively reviewed, rather than living forever in the background.
- Vendors and internal teams are held accountable for correcting flawed models and data, not hiding behind "that's how the system works."

## EMPLOYER ACCOUNTABILITIES

This Right is about every employer owning that its AI data streams are not infallible, and that they have an accountability to you to make things right, swiftly.

Data sources can be dirty, messy, stale, and ambiguous. Even when the data is correct, people change and grow, and your digital history must reflect that.

When an algorithm's judgment about you becomes your permanent label, it doesn't just misdescribe you; it reshapes your future. It can close off jobs you never get to apply for, training you're never offered, and promotions you're never considered for.

The right to swift correction is a refusal of that future. It demands that our tools reflect who we are becoming, not just who we once appeared to be.

# Right 5

## The Right to Psychological Safety and Mental Health, By Design

**No productivity is worth a broken mind, spirit, or soul.**

Pre-AI, we were already drowning in pressure, info overload, and burnout. We've lived through decades of ignoring the human science behind harmful cognitive loads.

Now AI radically ramps up those pressures. Relentlessly.

New research has found that **AI doesn't reduce workloads, it intensifies them.** In a UC Berkeley study, employees in AI-driven environments experienced greater workload creep, increased cognitive fatigue, burnout, and weakened decision-making than in pre-AI environments.[13]

This Right says:

- AI systems at work must not be built to hack your mind and actions for breaking-point output.
- They must be designed to support healthy, sustainable work that serves your life goals and happiness — not constant anxiety.
- Your emotional, mental, and physical health are not acceptable casualties of someone else's KPI.

### WHAT THIS MEANS IN PRACTICE

You have the right to have psychological safety, mental health, and guardrails preventing burnout **designed into your employer's infrastructure** — not tacked on as an afterthought in the form of wellness campaigns and programs, or just added to your manager's responsibilities. (See Part 2, Human Capital Operating System.)

### WHY THIS MATTERS NOW

**You are being hacked!** (If not already, you soon will be.)

Behavioral science and AI are coming together to create powerful new ways to shape and influence your behaviors. The same dopamine tricks that keep you doom-scrolling on social media can be used to keep you overworked.

As bestselling author Yuval Noah Harari cautions, "When we have enough data and computing power, we can hack human beings."[14]

At the center of people-hacking is a personalization-and-prediction loop. AI platforms study you, then select and rank content, notifications, and recommendations to trigger micro-reactions — curiosity, panic, fear, competitive urgency — that nudge you to keep doing things that are best for the company.

Signs of psychological harm by design include:

- Workers reporting chronic anxiety, sleep problems, burnout, or health issues tied directly to digital pressure.
- Cognitive Hacking: AI gamification hacks into your brain's learning machinery, leveraging reward prediction errors (RPEs) encoded in your midbrain through dopamine neurons. They fire more when outcomes are better for the company, delivering rewards that drive company-centric behaviors.
- Surveillance and Profiling: Every micro-interaction (like how long you viewed something sent to you by your manager) is logged and used to infer engagement, emotions, and vulnerabilities.
- Targets set entirely by algorithms that ignore human limits or task complexity.
- Notifications and nudges that keep you working longer hours, and make it harder to rest or to say no.
- Attention Capture and Self-Control Conflict: Behavioral addiction models that push you to be on your work apps more, check off more To Dos, and respond faster.
- Change Management moving from the soft-science of alignment, buy-in, and choosing to change into the engineered hacking of your dopamine reward system through your work tools.

As evidenced by the above, a machine that knows you is not always benevolent.

Understanding you is not the same as augmenting you. A machine that knows exactly who you are is just as likely to optimize for your weaknesses as it is to optimize for your growth.

## WHAT THIS RIGHT DEMANDS

- **Mental health as a digital design priority:** Measuring the cadence of work, workload notifications, and review of work rhythms — made easier to proactively track through AI. Human wellness is too often focused on after the damage has been done. If you truly care about people's health and wellbeing, you must design that into your digital infrastructure from the beginning.
- **Cognitive sovereignty:** Ban internal UX patterns that leverage cognitive hacking to hide healthy options (like rest or refusal) behind pressure, guilt, or confusion.
- **Humane targets:** Ensure that algorithmically-set goals are adjustable by human bosses and managers who understand context and capacity.
- **Opt-out without penalty:** Give workers ways to tone down or opt out of non-essential gamification and notifications — without being punished.

## WE'RE ALL PAYING TOO HIGH A PRICE

We've got to start treating burnout and constantly feeling overwhelmed as **structural violence, not individual weakness.**

Yes, we said it: **Structural violence.**[15] Which does not require malicious intent. Just designs that systemically prioritize company needs over their people's needs; that are continuously imbalanced in those priorities; along with a lack of leadership owning the consequences of their corporate-centered designs.

Stanford professor Jeffrey Pfeffer nailed this idea in his 2018 book: ***Dying for a Paycheck:*** *How Modern Management Harms Employee Health and Company Performance.*

And his insights were pre-AI. If we don't intervene swiftly and aggressively, look for lots more dying for paychecks, via the hacking of you.

# Right 6

## The Right to Human Judgment in Decisions That Impact Your Future

**A person, not an algorithm, must sign their name to your fate.**

When something big is on the line — your job, your pay, your hours, your reputation, a crucial decision — you have the right to be judged by a real human being, not only by a machine.

### WHAT THIS MEANS IN PRACTICE

People who know you, see you, hear you, believe in you, and understand you, must always be part of making decisions that affect you.

This Right says:

- For big decisions that can have a major impact on you — like being fired, demoted, or denied a critical opportunity — AI cannot be the final, unreviewed judge.
- A real human must be responsible for the outcome, and must be able to look at your specific story and all factors, not just your score.

### WHY THIS MATTERS NOW

When the wrong AI conclusions, based on flawed data, are applied to you, the results can be devastating.

There are already documented cases where workers were let go because an algorithm counted bathroom breaks and excused latenesses or absences as fireable offenses. This is moral malfeasance.

As many ethicists note, responsibility cannot be outsourced to code.

When career-altering decisions are automated, the harm can be deep and hard to repair — which is why frameworks like the EU AI Act classify many worker-management systems as "high-risk" and demand human oversight.[16]

High-stakes automation without oversight looks like:

- Automatic rejection of job applications with no human eyes on them, or no efforts to do additional screenings of the "Maybe" pile for criteria that AI may have missed.
- Automatic terminations triggered by metrics, with managers told, "This is policy."
- Disciplinary flags that follow workers around internally, with no way to clear their name.

If no one can step in, you are not being managed. You are being processed.

## WHAT THIS RIGHT DEMANDS

- **Humans in the loop:** For hiring, firing, major discipline, and harmful schedule changes, a named person must review and own the decision that impacts you.
- **Context matters:** Workers must be able to present information the system didn't see — illness, caregiving, harassment, technical glitches — and have it change the outcome.
- **No rubber stamps:** Reviewers must be empowered and trained to question and override AI outputs.
- **Records and review:** Track how often AI recommendations are changed by humans as one indicator that the oversight is real.
- **Far better criteria:** Automated tasks like scanning job applications are missing so much talent due to poor search criteria — missing amazing human skills like problem-solving, curiosity, creativity, grit, passion, and executive presence.

## SOMEONE MUST HAVE YOUR BACK

This Right is about proper AI governance, management, and assigning human-owned accountability.

Human oversight keeps responsibility where it belongs: With people who can be questioned, who can explain their reasoning, who must live with the consequences of their choices.

For decades, we've been searching for ways for automation to reduce people-management. Now is the time to consciously, deliberately, build it back up.

# Right 7

## The Right to Grow and Be Your Best In an AI-Transformed World

**AI should upgrade people, not make them obsolete.**

You have the right to keep learning and growing as AI changes work — not to be tossed aside as no longer valued.

Yes, **that responsibility begins with you.** Understanding what it takes to work smarter using AI. Retooling yourself at the same speed, or faster, than the ever-changing marketplace.

**But that's not all on you. Employers must partner with you.** It's their responsibility to ensure redesigned processes, tasks, and technologies are designed to leverage what humans do well, and AI cannot.

This Right says:

- As AI reinvents how we work, people must get real chances to learn new skills and move into better roles.
- Work should be redesigned so humans still use their brains, passions, and uniqueness — amplifying our amazing humanness.
- That redesign is a leadership responsibility. If leaders are putting upskilling all on you, or just leveraging AI for MoreFasterCheaper output from you, they are shirking their responsibilities and don't deserve you.
- Leaders must finally address what they've been avoiding for decades, long before AI appeared: They have an upskilling crisis, and must ensure upskilling keeps pace with constant disruption.

### WHAT THIS MEANS IN PRACTICE

You have the right to meaningful opportunities to learn, develop expertise, and grow your capabilities as AI transforms your role.

When tasks are automated, there must be a clear path for you to move into higher-value work, not shunted into a march toward obsolescence.

## WHY THIS MATTERS NOW

The World Economic Forum estimates that AI will disrupt hundreds of millions of jobs, and can also create new roles — if reskilling is prioritized.[17]

And yet the market keeps rewarding slash-and-burn strategies. Early in 2026, Block axed nearly half its workforce, 4,000+ jobs, based on the potential of AI. Hours later, its stock went up 25%.[18]

The difference between destructive disruption and strategic opportunity is whether organizations invest in human growth, or just in replacing humans.

The deeper we go into the AI Era, the more upskilling becomes an urgent priority.

Pilots who rely too heavily on autopilot lose manual flying skills, with deadly consequences when systems fail. Radiologists who let AI do all initial screening lose pattern-recognition expertise that catches AI errors, and patients can die. Writers who outsource all drafting to AI forget how to structure arguments or create emotional connections. Students who turn in AI-written papers as their own never develop critical thinking skills.

This pattern threatens every profession.

You know deskilling is happening when:

- Humans are expected to partner with AI, but are not given the necessary upskilling training, support, mentoring, and guidance to succeed.
- Automation plans are developed in secret, then announced alongside layoffs — no retraining, no transition paths.
- There is no protected time or budget for learning new tools or skills; training is an afterthought, done with no relief from daily and hourly pressures.
- Upskilling is limited to what's solely in the company's interest, not what increases your personal agency and success in a highly disruptive world.

## WHAT THIS RIGHT DEMANDS

- **Prioritize for meta-learning, future-proofing skills:** Task skills, tailored to a specific role using specific tools and doing things in a specific way, have an ever-decreasing half-life. Once machines became smarter than us, the only future-proof skills are foundational human capabilities, like critical thinking, resilience, sense-making, and decision-making. And, in no way does "take this mandatory course on how to use AI" count as real upskilling!

- **Proactive reskilling, upskilling:** Before automating major parts of roles, companies must analyze future human skill needs and offer real development opportunities and pathways.
- **Job redesign, not just job removal:** Roles should be re-crafted so humans handle the judgment, relationships, creativity, and complex coordination portions — with AI as an assistant.
- **Future-of-work redesigns:** AI is not just another tech tool to bolt onto existing processes. It forces major rethinks of why we work, how we work, and humans' role when machines can outperform us in many, but not all, ways.

## YOU DESERVE PATHS TO MASTERY

Deskilling, by starving people of real upskilling, isn't just an individual tragedy. It's an organizational catastrophe.

When your workforce loses or lacks future-proof competencies, companies cannot adapt fast enough when markets shift, AI systems fail, or when new problems emerge.

Work that offers no continuous learning, no new paths to mastery, becomes psychologically toxic. Humans need to feel they're developing, not declining.

Company approaches to AI that erodes that capability erodes meaning itself.

# Right 8

## The Right to Protect Your Time, Attention, Energy, and Inner Life

**24/7 data do not mean 24/7 ownership of your life.**

AI does not sleep. Computer servers don't have families. Algorithms don't get tired or need joy. But people do.

You have limited time on this planet, and you have the right and responsibility to prioritize your daily choices on what truly matters. To you... Not just to your company.

This Right says:

- Your employer must treat your time, attention, and energy as valuable assets that you are choosing to invest in them. You deserve better returns on those assets! This requires companies to stop all structural violence (page 23), and start delivering a lot more respect for those assets through worker-centered designs, controls, and experiences.
- You have the right to set boundaries: And AI tools must be designed to respect them.
- Productivity is personal, not just corporate. You have the right to expect company tools and processes to constantly make work easier and better and more efficient for you, not just them.

### WHY THIS MATTERS NOW

Digital tech has already blurred the line between work and life. AI takes it further.

With "smart" notifications, optimized for engagement and defaulting to corporate needs, predictive scheduling filling every gap, making "always on" feel normal and necessary.

The World Health Organization has linked long working hours to higher risk of heart disease and stroke[19], and Gallup estimates burnout costs the global economy hundreds of billions of dollars each year.[20] Humans need time to disconnect so their brains and bodies can reset.

Boundaries are under threat when:

- Performance reviews reward those who are always available and responsive, even during off-hours. And you feel guilty or pressured when declining always-on requests.
- Scheduling systems optimize you out of breaks and rest, treating unscheduled time as inefficiency.
- Senior execs create strategies for organizational efficiency, but push unresolved complexities onto middle- and frontline employees to sort out — resulting in constant fire drills, under-resourced demands, and unreasonable expectations.

## WHAT THIS RIGHT DEMANDS

- **Protection by design:** Respect for your time, energy, and attention must be built into every company work tool, every AI algorithm.
- **Clear norms with reasonable boundaries:** Written and practiced expectations about when workers are, and are not, expected to respond.
- **No stealth penalties:** Workers cannot be punished, formally or informally, for honoring agreed-upon boundaries.
- **Right to programmable tools and "off" switches:** Simple ways to turn off work notifications, requests, and monitoring outside of working hours. Simple ways to program, filter, and prioritize what comes at you, when, and how.

## YOU DESERVE BETTER RETURNS ON YOUR ASSETS

This is primarily a leadership issue.

Senior execs create strategies, then set schedules and budgets with no concept of what it actually takes to execute those ideas.

That cycle has been in place for decades: "Transformation" still means constant under-resourced fire drills.

While there are wonderful exceptions, as a whole, leaders have never held themselves accountable for the tremendous human toll of this approach:

- About seven to eight in ten employees across the globe are disengaged at work, suffering intense burnout, and costing the world economy hundreds of billions of dollars each year.[21]

This approach must end in the AI Era.

AI can help leaders anticipate and plan for real-world implementation challenges, create real-time feedback loops, and build human-centered solutions to almost any implementation problem. (Built into all ten Practices in Part 2.)

# Right 9

## The Right to a Safe and Valued Collective Voice in an AI-Shaped Workplace

**It must be safe to say, "What you're doing isn't working." And be heard. And valued.**

Workers on the front lines are the first to see when AI systems go wrong; when data or algorithms are biased, unsafe, unrealistic, or if the implementation plans have major gaps.

This Right says:

- Employee pushbacks and concerns are not an implementation threat to be managed; they are a safety system to be protected.
- People must be able to talk about how AI affects their work — without being tracked, profiled, or punished.
- AI cannot be used to predict who might organize, complain, or resist — and then silently bypass, strong-arm, or ignore them.
- Leadership must do a far better job than they have at seeking bottom-up feedback, honoring it, and incorporating it.

You should have real ways to give input, say "this is hurting us," or "this needs to be a lot better," and help shape and improve the tools and systems you are asked to use.

### WHY THIS MATTERS NOW

The AI Now Institute and many labor scholars argue that algorithmic management without a strong worker voice leads to brittle and unjust systems.[22] Workers are not just "end users;" they must be co-designers, safety inspectors, and moral witnesses.

But let's also appeal to every leader's profit and cost-driven focus:

> **Dear leaders, face it:** You currently suck at AI implementation!

Early attempts have failure rates reported as high as 95 percent.[23] You desperately need ground-level truths, insights, and ideas.

Stop trying to implement AI the way you've done everything else — driven by top-down plans with set-in-stone target dates and goals, and minimized investments in people. Start listening to your people!

Back to everyone who works... You have a problem when:

- There is no trusted way to report AI harms and poor implementation that is not controlled by the same teams that built the systems.
- Surveys and feedback are sanitized or watered down so senior leaders can keep doing what they've been doing, without confronting difficult ground truths.
- People who raise concerns about AI implementation plans mysteriously lose opportunities, get sidelined, or change-managed.

## WHAT THIS RIGHT DEMANDS

Start building valued feedback loops, with crucial bottom-up realities and insights. AI can make that easy to collect, synthesize, understand, and leverage.

- **Feedback with teeth:** Mechanisms where collective feedback can halt, modify, or reshape harmful or ineffective systems.
- **Workforce involvement:** Structures (councils, committees, unions, worker panels) that give workers a seat at the table when AI tools are chosen, designed, planned, and evaluated.
- **Safe reporting channels:** Independent channels for raising AI-related concerns, with clear non-retaliation commitments that are actually enforced.
- **A change in leadership behaviors:** Leaders valuing ground truths that indicate everything they haven't yet figured out or addressed.

## YOUR VOICE IS PART OF THE POWERSHIFT

This Right is about power distribution.

Throughout history, speaking truths to power has always been rife with barriers and futility. Now, with AI systems in the mix, which put technologists and algorithms in positions of power, this becomes even harder.

Leaping into the AI Era, and doing it well, requires more human-driven feedback and insight than most organizations are currently willing to seek, value, and leverage.

AI is terrible at understanding and addressing the politics, pressures, unexplored ground truths, and dysfunctional behaviors baked into transformations by all-too-human decision-makers.

We need real people providing difficult-to-hear feedback.

# Right 10

## The Right to a Share in the Value AI Creates from Your Work

**If you powered the gain, you deserve the upside.**

AI means fewer people can create more value — more sales, more products, more services, more profit.

But if all those gains go only to shareholders and executives, while workers get more pressure and less security, the system is both unjust and unstable.

### WHAT THIS MEANS IN PRACTICE

This Right says:

- You have the right to share fairly in the productivity gains, cost savings, and profits that AI generates from your labor and your data.
- People cannot be treated as temporary training wheels for machines that will then push them aside.

Company use of AI is different than, let's say, you working on a product where the company owns the patent and the rights to that product.

**AI isn't just mining your directed efforts on specific company-owned projects. It's mining YOU.** How you think, your passions, your feelings, your life experience. You! So, you deserve a share of that value.

That can come in the form of better compensation, improved benefits, reduced workload, greater job security, or deeper investment in your development.

Some thought leaders are even pushing for returns in the form of employee stock or company ownership. Or four-day workweeks, productivity dividends, or universal basic income. [24, 25, 26, 27, 28]

Gains from AI cannot flow only upward.

AI tech firms ran into this principle early on. They trained their foundational models on books, articles, research, and posts whose rights were owned by others. Class-action suits followed. In 2025, Anthropic agreed to a landmark $1.5 billion settlement for AI-driven copyright infringement.[29]

## WHY THIS MATTERS NOW

AI is a force multiplier for human effort.

When AI-driven efficiencies translate into higher profits or lower costs, but workers see only tighter quotas, more surveillance, and greater job insecurity, that's destructive.

Resentment grows. The social contract breaks. High performers leave. People quickly figure out that AI is being used to extract more from them, not to invest in them.

An unfair AI value split looks like:

- Major cost savings from automation, followed by layoffs and increased quotas for those who remain.
- AI tools that speed up work, but no upside change in pay, benefits, or workload expectations — just higher targets.
- Worker data is used to build valuable AI products, with no recognition or reward for those contributions.
- Leadership narratives about "AI transformation" that never mention how gains will be shared.

## WHAT THIS RIGHT DEMANDS

- **Fair distribution:** Mechanisms to translate AI gains into improved pay, working conditions, job security, and human development.
- **Upscaled investments in people:** Using part of the AI dividend to fund reskilling, better tools, and the redesign of work to be more humane.
- **Transparent accounting:** Honest communication about how AI has changed productivity, revenue, or margins — and who benefits.
- **Guardrails against pure extraction:** Policies that prevent AI from being used solely to cut headcount and squeeze remaining workers.

## WHERE'S YOUR AI DIVIDEND?

This Right addresses the central economic tension of the AI era.

Technology can create abundance, but if that abundance is concentrated in fewer hands while more people experience scarcity, it is destructive disruption.

Most of all, this Right recognizes a fundamental truth: AI does not create value alone.

It amplifies human effort, learns from human data, and succeeds because of human collaboration.

The value is co-created. The rewards should be too.

# If you remember one thing about these Rights, let it be this...

The future of work
is being rewritten now.

The only real question is
whether humanity
will be written into the code.

These are ten inalienable
human rights for the AI Era.

Everyone everywhere
deserves them.

Workers: Stand up for them.

Leaders: Stand behind them.

# Humanity + AI

**Most of the world is waiting outside AI's door.**

As of February 2026, ~ **6.8 billion people have never used AI** (lightest grey)

~ **1.3 billion** are using free (and outdated) versions of AI (dark grey)

~ **15-25 million** pay $20/month for the latest version of AI (7 mid-grey squares, bottom)

~ **2-5 million** use AI to write code (single black square)

**During the powershift, these Rights serve and protect all of us on this planet.** Including, and especially, the ~84% of us who have yet to use or experience AI —but will soon be very much impacted by it.

# OK.
# But...
# How?

# Part 2

## Transition Roadmap

### Ten Practices to Reimagine the Future of Work for Humans in the AI Era

**From Bill of Rights to Leadership Roadmap**

Part 1 drew the moral and design line in the sand.

Now comes the hardest part, the sustaining layer. Hard-wiring humanness and AI Era human rights into your entire infrastructure.

Creating a phased approach to diagnosing, piloting, scaling, and embedding human-centered practices into how your company operates.

**Building Your AI Era Human Capital Operating System™**

When humanity gets built into both people-driven and algorithm-driven operations, **that's your Human Capital OS**.

Think of it this way...

- The Bill of Rights is the **new rules of the game**; what's fair and what's right.
- The Human Capital Operating System is the **game board and the referee**, ensuring a balanced approach to market and company and workforce needs, equalizing their ability to succeed.
- **GenAI and AI agents are the power-up:** They make whatever game you built go faster and hit harder.

- **Your only real moat is your Human Capital OS.** Everyone is building AI-native structures. What they can't replicate is how you create Human+AI partnerships.

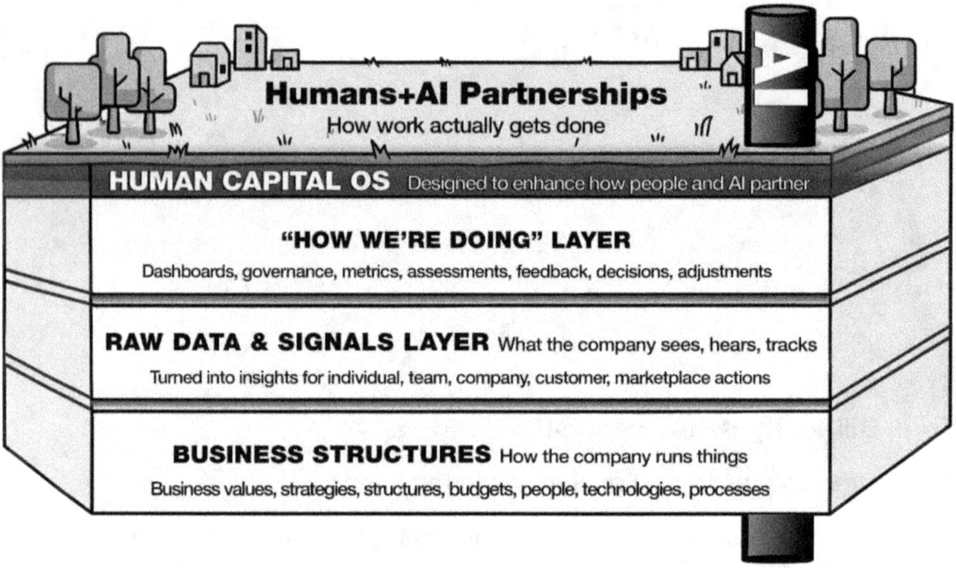


**In an AI-native organization:**
**Agentic and GenAI runs through every layer of the business.**

**The Human Capital Operating System** is specifically-designed to be human-centered, enhancing not only how people partner with AI to serve business needs, but also ensuring their AI Era rights are honored, and ensuring and enhancing their personal agency.
**Biggest challenge:** Mindset needed to build powershifts, not just taskshifts

**Enterprises adopt designs that help them win.** To help you succeed in implementing your Human Capital OS, we've built in RoFs:
**Return on Future** criteria are forward-looking AI metrics that track how effectively your investments prepare your organization for long-term advantages.

## Here's how we'll help you succeed in building your human-centered AI implementations...

**1. Scan this QR code**

**2.** Which takes you to www.CodeHumanity.ai, where you can **download a .pdf version of this book.** (Free with book purchase.)

**3. Upload the book's .pdf into your AI,** asking it to compare your company's structures, strategies, and goals to the ideas in the book.

**4.** You will get back detailed recommendations, **tailored to your organization's needs**.

WHY SEEKING TAILORED SOLUTIONS IS CRUCIAL

**You've got only a small window to completely transform your organization...**

# How Long You've Got

**MAXIMUM THRESHOLD for AI-NATIVE ENTERPRISE OPERATING MODELS**

While waves of job displacements will happen sooner, and the full diffusion of AI will play out over decades, leaders will likely experience a **tipping point within 5–8 years, (2.5–3 years in some markets),** where **AI-native enterprise operating models become** the new floor for competitiveness.

**KEY INFLECTION POINTS**

**2025 to 2026: Near-term Pivot:**[30] GenAI and Agentic AI take off, AI agents achieve teammate status.

**2026 to 2027: Crossing the Rubicon.**[31] AI investments drive major structural changes (replacing employees, AI agents outnumbering humans in some functions). "Frontier AI" for enterprises shapes the future of work.[32, 33, 34 35]

**2031–2034**: **Maximum Tipping Point Threshold.**[36] No one knows the exact date, but... Given current AI trends and leadership's AI investments, views, and strategies, ROI-driven AI automation and operating model changes will be in place.[37, 38] And that's excluding AGI, likely here by early 2030s![39]

**Tipping Point Challenge:** Run your own assessment. Whatever timeframe you come up with, **own the urgency!**

# How Long Your Ancestors Had

### INDUSTRIAL REVOLUTION

~ **1870:** Approximate mass-use anchor that includes major companies and consumers

~ **Tipping point, 1920 to 1930:** Approximate timeframe when new enterprise operating models were mainstreamed

**Timespan:** ~ 50 to 60 years

### DIGITAL REVOLUTION

~ **1989:** Approximate mass-use anchor

**1990s to early 2000s:** Infotech, the Internet, email, smartphones, ERP, reengineering core processes. Operating models were radically disrupted, becoming digitally-native.

**Tipping point, 2005 to 2010:** Cloud-based Big Data, CRM, ERP truly reimagined, and locked-in, digitally-driven company operating models.

**Timespan:** ~ 20 years

### THREE DIFFERENT ERAS OF OPERATING MODEL REINVENTIONS

We start the clock when the new technology reaches everyday use for major companies and at least 20% of households.[40, 41, 42]

We then mark the point when a clear way of managing and organizing around that technology becomes the norm in large firms.

The timespan between those points is **the practical window leaders have to redesign their organizations** before that new model becomes the new standard.

# This is the story of how leaders designed Human Capital Operating Systems.

You are creating the future's
compelling narratives.

With these ten Practices,
you will create the stories
that will be told for generations.

Of how you fought for our souls,
battled for us humans...

How you created powershifts,
and a future that made us all proud
to be part of the AI Generation.

Or how you didn't.

The choices you make
in the next five to eight years
determine how this transition
will be remembered.

# Human OS Practice 1

## We Need AI Era Leadership

We Need Better Leaders, Better Leadership

### MINDSET, WHY WE MUST CHANGE

Leaders across the globe are realizing that AI transformations are completely different than anything they've ever experienced before. When stepping down in 2026, the CEOs of both Coca-Cola and Walmart proclaimed exactly that. "I could start this transformation," said Walmart's Doug McMillon, "but I couldn't finish it. It's very clear that AI is going to change literally every job."[43]

**The trillion-dollar blind spot:** That's what Peter Diamandis, founder of the XPrize Foundation, called it.[44] The real question isn't "How do I integrate AI into my business model?" He said the driving question needs to be: "**What happens when my customers become smarter than my company?**"

You need to move from treating AI as yet another technological disruption to mastering how to quickly own every new market cycle.

Intelligence, and a lot of what used to make your business special, is now commoditized. This forces a complete rethink and reimagining of everything.

**Some of the leadership truths AI forces us to face:**

- AI exposes where we never redesigned work in the first place.
- It is a **force multiplier** for leadership strengths and flaws, strategies and systems. It makes visible fragilities and assumptions that were always there but hidden.
- In doing so, AI is accelerating the gap between how leaders believe work gets done and how it actually gets done. If we don't redesign decision rights, incentives, and accountability structures alongside the tech, we're just pushing friction downstream.

## ARE YOU LEADING A TASKSHIFT OR A POWERSHIFT?

You are leading a true AI-driven powershift when at least five key criteria are met...

1. **Disrupt and reshape the strategic debate.** Taskshifters push solely for "free" AI teammates to cut costs. You will push for rethinking business models, structural changes in how power, judgment, controls, governance, and value creation are distributed. You'll be focused on reimagining the future of Human+AI work.

2. With every AI move, among the first questions you must ask and address are: **"Within our company, whose agency increases? And whose risk increases?"** (There will be winners. And losers. Name them.) Be specific about how and why your AI strategies will impact them. Be clear about the strategic choices you are making for those who will be negatively impacted. Own that impact.

3. **Workforce voices are leveraged as true co-authors** of AI strategies, not just for feedback on implementation details after critical decisions have already been solidified.

4. **You set locked-in human rights constraints** on AI before budgets and vendors are approved.

5. Board reviews include: **"How does this AI design erode or strengthen our Human Capital Operating System?"**

**AI is the mirror and amplifier, not the actor.** It reflects your organizational designs, leadership priorities, talent choices, values, and processes.

When you build systems that treat humans as components to optimize, AI will optimize them into exhaustion and extinction, or sack them.

For example: In their late 2025 study of over 1,000 global executives, Tom Davenport and Laks Srinivasan found that leaders were partly to blame for traumatizing the workforce: Companies are laying off workers because of AI's potential, not its actual performance. "The job losses and slowed hiring are real, even though companies are waiting for AI to deliver on its promises."[45]

This happened to Swedish fintech giant, Klarna, one of the loudest voices championing AI in customer operations. In 2022, they laid off 700 people, only to have customer satisfaction tank, and, within two years, they had to refill many of those positions with humans again.[46]

**In the age of Human+AI partnerships, leaders are gardeners:** Designing conditions where people, teams, and technology can thrive together without burning out or breaking trust.

That means changing how we seek, select, and evaluate AI Era leaders, as well as the strategies they approve and build. Which is the foundational work essential for successful AI implementation.

"We started focusing on EQ a couple years ago as part of our leadership development programs," says Ravi Jayasekera, Chief People Officer at Hemas Holdings. "Focusing on inner work, introspection. Because most humans are unaware of their inner selves; why they behave certain ways, how they make decisions, and their motivations. If a leader has not done that inner work, they can't get the best out of their teams or themselves in an AI-enabled workplace."

## SYSTEM RULES AND ARTIFACTS, WHAT MUST EXIST
### Translating Digital Transformation into People Transformation

- **Redefine success beyond productivity and profit.** Add human sustainability indicators such as learning velocity, psychological safety, agency, and long-term employability. Track these with the same seriousness as profitability, costs, and EBITDA. If you only measure output, you will only get output — until people break.

- **Run "living systems audits" alongside financial audits.** Besides financial targets, quarterly reviews should ask: Are our people more capable than last quarter? More resilient? More connected? Or are we mining them like a depleting resource? Track capability half-life, not just productivity curves.

- **Design work systems that regenerate energy.** Build in recovery time, limit cognitive load, rotate high-intensity work, and treat rest as productive. Work should return people to themselves stronger than it found them.

- **Commit to earned trust.** No secret monitoring, clear rules on data use, and easy ways to correct wrong information.

- **Shift from "managing performance" to "stewarding capability."** Performance management implies control over outputs. Stewardship implies responsibility for growth, health, and future potential.

- **Build participatory dashboards.** Let teams see and challenge their own data in plain language: "What we measure, why, and what we will never measure." Use AI to reinvent performance measurement so it illuminates contributions and system health, not just volume.

- **Increase organizational sense-making capacity.** Here, AI can be revolutionary:

    - **Worker Impact Assessments before every AI deployment:** Plain-language reviews of workload, dignity, bias, privacy, and job quality, with employee representatives empowered to hit "Pause" if risks are not mitigated.

    - **Algorithmic Justice Annual Report:** Transparent audit of AI people decisions (hiring, pay, scheduling), co-reviewed by employee panels.

    - **AI Tool Registry and Decision Log:** Live list of every AI system touching workers — data used, outputs, owners, appeal paths — ending "no one knows how it works" excuses.

    - **A governance layer** that enforces "human must approve" rules for high-stakes calls and logs when AI decisions are questioned or changed.

- **Re-examine leadership principles.** With AI automating routine work, redefine principles around stewardship, not command — because black-box leaders create black-box AI systems.

- Rewrite **leadership job descriptions** to include responsibility for system health — well-being, learning, inclusion, ethical tech use — not just revenue or cost targets.
- Establish **leadership covenants** that spell out commitments to explainability, consent, contestability, and psychological safety.
- Provide **AI literacy and reflection tools** so leaders can see how their choices shape patterns in workload, attrition, and equity.

James Scott, an investor in digital fintech firms and board member on several companies, pinpoints one of the biggest challenges in implementing all of the above. "In large organizations, budgets are not the issue. Mostly, leaders see themselves as **massively constrained by their own infrastructure** — their previous commitments to data flows, and their compliance and risk functions. Reimagining and rebuilding data flows to properly leverage AI is often their biggest barrier."

Yes, rebuilding for AI is a huge undertaking! But Scott stresses, leadership's mindset shift comes first. "I spend every day focused on businesses that are going through major AI and digital transformations. **100% of them do not have the human-centered foundations to be ready for AI transformation.**"

## TECH STACK: WHAT SUPPORTS THIS

The Human Capital OS runs out of the semantic layer, a business-friendly abstraction layer that sits between complex data sources and end-user tools.

Deploying integrated people analytics platforms that combine traditional HR metrics with human-sustainability indicators, such as burnout-risk signals, capability-growth tracking, psychological safety pulses, and workload-balancing systems.

These are not surveillance tools; they are diagnostic instruments that reveal whether your system is regenerative or extractive. Think of them as organizational MRIs that show system health, not individual deficiencies.

Key elements include:

- A secure people-data platform that integrates HR, collaboration, learning, and operational signals without exposing individuals unnecessarily.
- Simple analytics and visualization tools that non-analysts can read and question.
- Governance rules that define who can see what, for what purpose, and how long data is kept.
- Tools that log AI-influenced decisions, flag risks in real time, and provide dashboards for leaders and workers to see and contest outputs — with "human-must-approve" rules on high-stakes calls.

## THE PHILOSOPHICAL GROUND

In Gallup's 2026 report, *The World's Most Important Problem*, employment challenges are among the top four. "Simply creating more jobs addresses only part of the problem," says the report. **"Dignified and engaging work** that maximizes people's strengths is often elusive to most workers worldwide. **Progress toward this goal benefits individuals, businesses, and economies alike."**[47]

AI Era workplace transformations will be the biggest leadership challenge in generations.

When we shared all ten Practices with the head of Data, AI, and Analytics at a Fortune 50 company, they responded, "**The biggest transformations will come in reimagining workflows.** That's why these practices are so important. To leverage them, organizations will need to rethink entire decision structures. Instead of improving processes, leaders need to ask, 'With AI, do we need these roles or processes at all? This will lead to completely redesigned organizational structures and governance."

One of our most wide-ranging conversations was with Ram Srinivasan, Managing Director for Future of Work and AI Adoption with JLL Americas, as his role is to help leaders from all industries prepare for these ten Practices. Some of his wisdom about leadership in the AI Era...

## HUMAN CAPITAL OS: YOUR ONLY SUSTAINABLE MOAT

**You have five years—eight** at most, and as little as **2.5 to 3 years** in some markets— before competitors deploy their fully integrated, AI-native enterprise operating models.

After that tipping point, technological parity erases differentiation. Your edge then becomes how brilliantly you pair human capability with machines to deliver unique value.

Enter your Human Capital Operating System: **Infrastructure that could not have been designed or built before 2026,** it's your company's bespoke blend of reimagined work, people, and AI—**engineered specifically to amplify the Human + AI relationship.**

Where tech once shaped many behaviors, **the Human Capital OS makes human needs an equal design force**. I/O psychology, learning and upskilling, and emotional safety matter as much as performance management, governance, and budgets. (See Practice 2, Taijitu Design Principles.)

It's an intentionally designed space for humans and AI to co-create and thrive together in ways and at scales neither could achieve alone.

Your uniquely designed Human Capital OS is your only sustainable protective moat.

- "**You will be leading a century's worth of innovation within the next decade!** You'll have to think critically about new ways to create value, new ways of working, new ways of thinking."

- "IDC projects 1.3 billion AI agents in workplaces. That's a gross underestimation. AI agents are like slides. How many slides do you have? Do you count them? You toss them once they've served their purpose, right? That's how agents will be. Agents will be the countless bridges between your human teammates and company goals and structures. But leaders have never had to think about this before in how they build work ecosystems."

- **"A mindset shift has to happen within every level of the organization.** Nobody ever said that the organization structures we've been using for a couple of hundred years were meant to last forever."

- "We're going to invent completely new categories of work. Who knew that a company would pay $300,000 to hire a storyteller? This is already happening! The people side of work is going to expand in many dimensions."

- "Too often we confuse our leadership role with our job. The role that I play — being there for people when they're in trouble — isn't in my title, or on my LinkedIn profile, but it's part of the truth that I live. Leaders creating AI transformations would do well to think of the Hindu concept of Dharma, or duty. Yes, you have a duty to your company's financial success. But **your role in this disruptive transformation is also your duty to society, your family, and all your teammates."**

## MONDAY MORNING STARTER KIT FOR C-SUITE, LEADERS

**Do not start with technology or use cases. Do not start with ROIs.**

Begin where all strong AI Era leaders must begin: With the work itself, and the people who do that work. Just like the TV show, Undercover Boss.

- Every senior team member: **Do someone else's job for a day,** as close to the frontlines as possible. Learn the ground-truth struggles built into your current systems and structures.

- Have deep, vulnerable conversations with your people — not just about AI and work, but about them; their struggles, dreams, goals, and needs.

- Come back together and process what you learned: About your people, how remarkable they are, and what they need to build Human+AI partnerships.

## BACK FROM THE FUTURE WITH GEN ALPHA

The tail end of Gen Z (born 1997-2010) and Gen Alpha (born 2011-2025) will be the first AI Native generations. Because what matters to them should matter to you, we conducted a gathering of two dozen teenagers, 12 to 19, to discover what's on their minds, and paired them with dozens more industry mentors and technologists for a two-day sprint on all things AI. Results from that Techathon...

**Warnings From the Future**

• Companies should help reskill workers if AI replaces parts of their job (83%)

• People should be able to control or delete their personal data (87%)

• AI will create equal amounts of problems for me as it will opportunities (71%)

Most crucial to how you lead were their responses to...

**What Adults Should Get Right Before AI Shapes My Future**

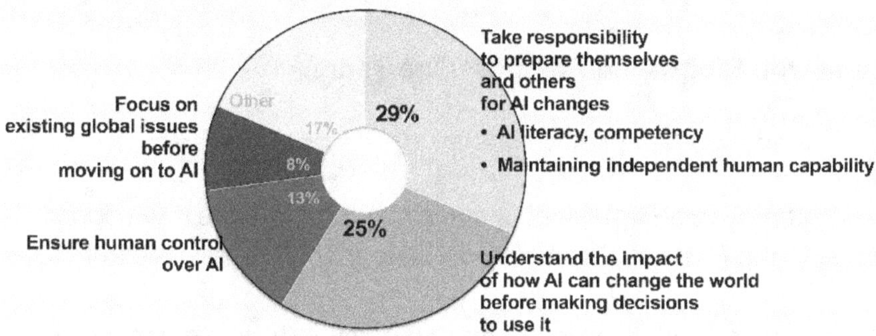


Our sampling from the future is **holding you accountable for the future they'll have to live in.**

This includes your personal use of your organization's AI tools. "Most leaders form opinions about AI without truly understanding it," one digital transformation leader told us. "It's not like PowerPoint that you can casually 'pick up.' It has vast, wide-ranging capability. If you want to lead AI transformations properly, you have to study it, experiment with it."

### Then Use This Book to Revolutionize Your AI Strategies

**Scan the QR code on page 51. Upload the book's .pdf into your AI:** Within your prompt, describe your top three AI transformation priorities, as well as what you learned from your Undercover exercise.

The results will likely be both reassuring and jarring. **1.** Some strategies you've already begun likely need just minor tweaks. Congrats! **2.** However, most everything related to powershift-not-taskshift is likely to be a major mindset shift for you and your organization.

*That's* where this book will be most helpful.

### Prioritize, and Build from There

The work ahead will be highly disruptive and often difficult.

To simplify it: **Do not approach the ten Practices as one massive transformation effort.** At least not yet. Too much. Too complex. Too easy to fail.

Start with **One Right, One Practice, One Team.**

Picture this as a 90-day pilot....

- You pick **one Right** that is already hurting people if you ignore it — say, the Right to Psychological Safety and Mental Health by Design.

- You pair it with **one Practice** that gives you real levers — say, redesigning the workforce journey around agency and well-being.

- On Day One, choose **one team** that feels the pressure every day: A customer support pod, a call center squad, or a product team overloaded with back-to-back meetings.

- You sit with them. Asking about their focus, their sense of control, and what they need to work smarter, not harder. Then you make one clear promise: "For 90 days, this team will be a testbed for work that protects mental health while still hitting real goals."

**Over those 90 days,** you change a few concrete things, not everything. Like building small, visible guardrails into the team's rhythms: Including a simple check-in at the end of each sprint on the team's energy and focus — not just output.

**By month three,** you extend the lens beyond that squad. You codify what worked into your Human Capital Operating System: The way you plan capacity, design journeys, and govern AI nudges. You are proving, one team at a time, that AI can amplify humanness instead of eroding it, and then baking those proofs into the way your whole organization works.

This approach helps name many of the AI Era tensions you're already facing so you can better address them.

Launching pilots of one Right, one Practice, with at least one team or function, will help you expose and address system weaknesses or challenges, early on.

In a world where every company can buy the same AI tools, **how you design your Human Capital Operating System is the only durable advantage you have.** Remembering, as with all disruptive changes, AI adoption is emotional first.

### AI Implementation Has Nothing to Do with Technology

**It's all about leadership.**

That was the consistent thread we found during interviews for this book.

One organization **introduced AI to analyze employee interaction patterns** to better understand collaboration and career progression. It was met with a wall of resistance. Seen as surveillance. So leadership relaunched the effort by fully explaining the pilot's purpose; clarified how the data would benefit employees, and ensured transparency on how insights would be used. Result: Resistance was greatly reduced, employees appreciated the goals, and the company gained important insights.

Another company leapt over those missteps: They launched AI initiatives by **shifting from a top-down rollout to co-creation with employees.** They involved employees early on; invited their collaboration on governance creation; and maintained continuous dialogues on the impact of AI on people. Result: Higher engagement, reduced resistance, stronger employee ownership.

Another organization observed employees becoming quiet and **disengaging after AI's launch.** So they normalized the questioning of AI implementation plans, rewarded early error detection, and separated AI outputs from final decisions. Result: Increased engagement, stronger team trust, improved decision quality.

Finally, **if you're not truly leading, you'll get Tokenmaxxed.** Your employees will quickly figure out the holes in your AI productivity strategies, and game your system. "Tokenmaxxing"[48] is already happening. That's where employees max out unimportant uses of AI simply because their employer is tying performance reviews to how much they're using AI tools.

### One Crucial Solution to Better AI Era Leadership

**More women leaders!** Recent research by Chief/Harris Poll found that 87% of women leaders polled have witnessed, and are concerned about, the negative consequences of prioritizing AI without parallel investments in people. And 85% are creating strategies for human skill development, establishing AI Era governance guidelines, and are actively working on what good judgment looks like in the human-agentic workforce. They're also asking harder questions about the consequences of moving too fast without considering the human costs.[49]

## ORGANIZATIONAL KEY METRICS

Do You Have AI Transformational Leaders?

**Process Metrics** Signals leadership shifts into the AI Era

- Frequency and quality of living-systems audits are included alongside financial reviews (psychological safety, learning velocity, burnout risk, and agency scores for major units).
- Percentage of AI initiatives that include Worker Impact Assessments and employee co-design sessions *before* go-live.
- Proportion of leadership scorecards that weigh human-sustainability indicators as heavily as output and EBITDA.
- Number of leaders whose job descriptions and reviews explicitly include stewardship of the Human Capital Operating System.

**Business Results** Outcomes tied to AI Era leadership behaviors

- **Increase Return on Future (RoF)** on AI investments by 20–35% within 2–3 years as leadership shifts from tool deployment to stewarding an integrated Human Capital Operating System, reducing pilot waste and concentrating spend on scalable, reusable AI+People capabilities.
- **Cut strategic execution risk** on major AI and transformation programs by 25–40% over the next 24–36 months, as leaders systematically use living-systems audits, Worker Impact Assessments, and participatory dashboards to surface human-capital fragilities before they derail large bets.
- **Accelerate the organization's capacity to launch new AI-enabled** products, services, or business models by 25–50% in 2–3 years, driven by an "AI Era–fit" leadership bench that is fluent in Human+AI work design, distributes decision rights closer to the work, and can rapidly reconfigure teams and structures without eroding trust or agency.
- **Strengthen leadership accountability** as executive scorecards explicitly track human sustainability indicators with the same rigor as revenue, cost, and EBITDA.

**Individual Leader Key Metrics** Tracking behavioral changes

Each leader can track:

- Days per quarter spent in their people's reality: Doing the work, shadowing, or holding open, vulnerable conversations.
- The ratio of time spent fixing systems versus "fixing" individuals (changing workflows, load, and rules versus dealing with performance issues).
- Whether human-sustainability indicators (learning, safety, energy, agency), not just output, are trending up over 12–18 months,
- How their people would answer, anonymously: "Do I trust this leader to have my back in the AI Era?"

**MONDAY MORNING STARTER KIT** **FOR INDIVIDUAL CONTRIBUTORS**

AI Era leadership is not only those who sit in powerful positions. It is also about **self-leadership** and personal growth.

You have the **right** to human agency in an algorithmic world; and the **accountability** to upskill and be your best, no matter what comes at you. Each of the ten Rights carries new personal responsibilities.

**Uncomfortable truth:** AI Era leadership also begins with you. Place to start: How many of the seven superpowers in Practice 8 have you mastered?

You can use AI to power-up your game. Uber employees recently built an AI clone of their CEO to prepare presentations to him. The clone helped them refine their slides and sharpen their pitches before the actual meeting.[50] You can reshape internal corporate dynamics from the bottom-up, using AI.

## CLOSING REMINDER

AI is a force-multiplying mirror, not the actor. It reflects back to you — and everyone — your strengths and flaws. And, it's a powershift, not a taskshift. Once we embrace that, **AI is your lever to move the world!**

# Human OS Practice 2

## People Analytics Without Fear

From Surveillance to Shared Experiences and Insights

### MINDSET, WHY WE MUST CHANGE

**Everything in the new People Analytics will always be about trust.** How well you build a trusting relationship with your workforce.

Your workforce is asking: "Can I trust you to have my back, and care about my needs and concerns, as you build these AI and analytics systems? When building human and AI partnerships, am I being valued? Will my personal data, like me tracking my own goals and development, be protected as confidential?"

When people trust how data is being used, they use it to improve. When they fear it, or can't see or understand it, they game the system, check out, or leave.

**The Panopticon Temptation:** AI makes it possible to track everything about everyone: Productivity patterns, communication networks, sentiment trends, flight risk scores, engagement levels, and collaboration indexes. We can now turn organizations into total information environments where every individual action generates data.

**The Radical Alternative:** Use analytics to diagnose your Human Capital Operating System. Ask: "Are our systems, tools, processes, and dashboards revealing patterns in work design, team health, and organizational systems that everyone can see, discuss, and leverage at their level?"

The goal is better data-driven work systems, that serve people, teams, and the organization. Seeing and solving for the unglamorous frontline constraints that have never before shown up on organizational dashboards, but define the employee experience.

Those experiences must be built on shared trust. “AI is not just a new technology; it is an entirely new system of thought processing and problem-solving,” said Errol Price, Legal Director for Symmetra, whose focus is on workplace law, human rights, and AI governance. “Roles will change, expectations will evolve, and organizations must be open about what is known and unknown. At the broader human level, **success means that we preserve dignity. The biggest loss would be the undermining of trust.”**

## SYSTEM RULES AND ARTIFACTS, WHAT MUST EXIST

- **Data for monitoring systems, not surveillance of individuals.** Aggregate data reveals patterns in workloads, burnout risks, skill gaps, and team health — without exposing personal details unless explicitly consented. No “gotcha” dashboards.
- **Transparency as the default.** Every metric explains itself in plain language: What it measures, why it matters, how it’s calculated, and what actions it suggests. No black boxes.
- **Right to contest and correct.** Workers can challenge data about them and see changes made in real time. Errors don’t haunt careers.
- **Participatory sensemaking.** Data reviews are collaborative sessions where teams interpret findings and co-design fixes, not top-down pronouncements.
- **Human-centered metrics first.** Track learning velocity, psychological safety, agency, energy regeneration, and equity alongside output and costs. Balance human sustainability with business results.
- **AI decision logs.** Every AI-influenced people-decision (hiring, scheduling, performance) is logged with explanations, overrides, and appeals — visible to affected workers.

“Because people analytics and performance management require contextual and ethical judgment, over-reliance on automation is risky,” says Kumudu Dias, Director of Finance and Control at Nestlé. “During early stages of implementing AI, companies will find that they are still dealing with inaccurate, dirty, and incomplete data. And we know that AI can be confidently wrong. The best approach is a human-in-the-loop model. AI should support KPI standardization, data analysis, and calibration, while humans lead in evaluating leadership, ethics, and final decisions.”

## REIMAGINING PEOPLE ANALYTICS: FIVE MUST-DOs

If people analytics are to help both the company and each individual succeed and thrive, your systems must do this...

1. **Work backward from the worker benefit.** Every new metric or tracking by AI must also be attached to a worker benefit statement: **"Here's how this will help you do better or live a better life."** If you can't do that, you're not ready to collect the data.

2. **Make data subjects co-designers.** Workers, not just executives and vendors, **must help decide** what gets measured, what's off-limits, and how insights are shared back. You treat their lived experience within your system as design input, not noise or resistance.

3. **Governance that protects dignity.** Create, publish, and stick by strict rules on what data is collected, who sees it, how long it's kept, and what's off-limits (emotions, biometrics, off-hours activities). Independent audits enforce compliance.

4. **Separate support and punishment.** Distinguish between **data for coaching and learning** vs. **data for discipline and termination.** Do not quietly repurpose one for the other. People know which is which.

5. **Give people access to their own story.** Create and deliver simple ways for the workforce to **see, question, correct,** and do scenario planning for improvements with data attached to them. Treat this as part of their rights, not as a favor.

**If these five are missing,** your people analytics strategy is not a strategy. It's an extraction machine with a prettier name.

## TECH STACK, WHAT SUPPORTS THIS

Build integrated platforms that combine HR, collaboration, learning, and wellbeing signals into system-level insights, not individual surveillance.

Core components:

- **Secure data lake** for people signals, with strict access controls and anonymization by default.
- **Plain-language dashboards** everyone can read, question, and contribute to — no Ph.D. required.
- **Real-time contestation tools** so workers can flag errors, add context, and see updates.
- **Sensemaking AI** that surfaces patterns ("Workload spiking in Q4") and suggests fixes ("Rotate high-intensity roles").
- **Governance engine** that enforces rules, logs decisions, and flags risks (biases, privacy breaches, etc.).

Examples: Workday People Analytics with custom human-rights overlays, or Microsoft Viva Insights reoriented to include human-centered system health.

## PHILOSOPHICAL GROUND

People Analytics needs a new design principle driving every data collection, every dashboard, every data-driven decision: **Taijitu Principles** (tie•gee•too, also known as Yin-Yang).

This means that your commitment to enhance each individual's ability to create and control their own destiny **must be equal to and balanced with** ensuring that the company can create and control its future. One ensures the other in a continuous Yin-Yang loop.

For example, "When it comes to how our people are using AI, we focus on **two sets of KPIs,"** says Kosala Hewamadduma, Global Marketing Director at Flora Food Group. "Those who **use** AI. And those who **lead** with AI. 'Using AI' means you asked a model for content. 'Leading with AI' means you are redesigning how your organization listens, creates, scales, and learns with governance and brand integrity built in. One is tactical. The other is transformational. One can be copied overnight. The other becomes competitive advantage."

You are accountable for enhancing corporate success and individual success, in equal measure. By design. The leveraging of all people data serves two stakeholders' needs — equally.

**MONDAY MORNING STARTER KIT FOR C-SUITE, LEADERS**

**Week 1: Audit your current people data.**

- List every source (HRIS, email analytics, collaboration tools, etc.).
- Map what it measures, who accesses it, and how it's used.
- Rate each on transparency, consent, and human-rights alignment (1–5).

**Week 2: Run a trust pulse.**

- Anonymous survey: "Do you trust how we use people data? Why/why not?"
- Hold sensemaking sessions with teams to hear unfiltered truths.

**Week 3: Prototype one regenerative metric.**

- Pick a team, track learning velocity or energy regeneration for two weeks.
- Share findings openly, co-design one fix, measure impact.

**What you gain:** Proof that trusted data drives better decisions and engagement.

## WORKER DATA BILL OF RIGHTS

Embed these as non-negotiable governance rules:

- **Right to Know:** What data, why collected, who sees it
- **Right to Contest:** Challenge inaccuracies with guaranteed review
- **Right to Forget:** Delete personal data when no longer needed
- **Right to Opt-Out:** Decline non-essential tracking without penalty
- **Right to Aggregate Only:** No individual profiling without explicit consent

## ORGANIZATIONAL KEY METRICS

Is Your People Analytics Regenerative?

### Process Metrics

- Percentage of people data sources with full transparency documentation (≥90%).
- Number of data contestations filed and resolution rate (≥95% within 48 hours).
- Frequency of team sensemaking sessions on data insights (monthly minimum).
- Compliance with Worker Data Bill of Rights (100% audited systems).

### Business Results

- **Increase Return on Future (RoF)** of AI and analytics spend by 20–30% within 2–3 years as people data is redirected from individual surveillance to system-level insight, enabling leaders to prioritize investments that build reusable Human+AI capabilities (learning velocity, adaptability, inclusion) instead of one-off monitoring tools.

- **Reduce the failure rate of future AI and analytics initiatives** by 30–50% over the next 24–36 months through trusted, transparent people analytics that expose workload, capability, and culture risks early — before they undermine large-scale automation, restructuring, or new business models.
- **Shorten time-to-value for new AI use cases** by 25–40% in the next 2–3 years as a high-trust, contestable people-data infrastructure allows faster experimentation with Human+AI workflows, cleaner feedback loops from the workforce, and rapid scaling of what works across functions and geographies.
- **Increase ROI on HR and AI investments** by 25–30% as transparent dashboards reveal which processes, leaders, and teams actually create value.
- **Convert people analytics from a monitoring cost center** into a decision engine, enabling business units to unlock 10–20% productivity gains without higher burnout or attrition.

**Analytics Team Member Metrics** Tracking strategic impact

- Ability to answer: "What question is this data answering, and how does it help improve the system rather than just watch individuals?"
- Are you having open, ongoing, honest bottom-up performance and development conversations with the people who actually do the work that's being assessed?
- Feedback from individual contributors and teams: "Does this data help us improve?" (≥80% Yes)
- Feedback from senior leaders: "Does this data help us manage results?" (≥80% Yes)

**MONDAY MORNING STARTER KIT** **FOR INDIVIDUAL CONTRIBUTORS**

You, or teammates who represent your views, should be invited to be **co-designers** of the new People Analytics.

**Uncomfortable truth:** If you do not see major leaps of progress on the ideas laid out in Practice 2 within the next couple of years, you should consider changing employers.

**Individual key metric:** Evaluate how courageously your leaders have reimagined their People Analytics system. **If they are courageous,** these are the kind of survey questions or themes for corporate-wide conversations and data collections you should be seeing:

- "Can you achieve your personal dreams and goals by working here?"
- "Are we providing the reskilling, upskilling, and personal agility support you need to succeed, thrive, and grow in the AI Era?"
- "Do you believe our leaders have your back as they lead us into the AI Era?"
- "Has our use of AI made it easier for you to work smarter, not harder? ...Saved you time each day? ...Reduced your aggravations and effort?"

## CLOSING REMINDER

AI Era People Analytics is totally about generating **greater, deeper, and more profound trust** — between people, in all data, in dashboards, and in all Human+AI partnerships. Moving from surveillance to shared experiences and insights, so that everyone, not just the company, can thrive, succeed, and create and control their own destiny.

# Human OS Practice 3

## Algorithmic Justice and Worker Safeguards

### MINDSET, WHY WE MUST CHANGE

AI doesn't just automate decisions. It silently moves power.

**When an algorithm is your boss, it needs a boss too.**

If no one can pull the plug on an AI system from churning out challenged information, you don't have governance; you have a god. Algorithmic justice brings invisible power out into the open and makes it answerable to human values.

Stated simply, algorithmic justice means that every AI decision impacting an employee's future must be explainable, contestable, and auditable, by the people it affects. It gives everyone easier ways to access, understand, and participate in AI-driven powershifts.

### SYSTEM RULES AND ARTIFACTS, WHAT MUST EXIST

**Building a Trusted System Through Algorithmic Justice**

- **Live register** of every AI system touching workers: What it does, data used, owner, decisions influenced.
- **Require worker impact assessments and Stop/Go gates.** Evaluate workload, dignity, bias, and privacy before rollout. Worker reps can hit Pause until necessary mitigations are created.
- **Fix it fast.** Time-box investigations, name owners, and record fixes publicly. Normalize reversal and apology.
- **Ethics built into KPIs.** Inclusion, agency, and well-being on leadership scorecards.
- **Algorithmic Justice Panel** with worker reps empowered to reshape deployments for enhanced implementation.
- **Plain-language impact assessments** pre-rollout.

## JUSTICE FOR ALL: THREE MUST-DOs

To build true worker safeguards and to deliver algorithmic justice, you must commit to...

1. **No more black-box tracking: Radical transparency.** Understand that every algorithm and data stream into it is about one-thing: **Your relationship with your employees.** And trust cannot exist between the parties until you commit that they get to see the AI tracking that's attached to anything they do.

2. **Assign a human owner for every AI data-stream tracked.** Each major AI-at-work tracking effort has a named leader, accountable for its impact on people, its ethical foundations, and its impact on performance — not just its technical upkeep and reports.
3. **Justice is founded on participation and involvement.** It is designed and fine-tuned with the help of the people who are impacted by it. Workers must be involved in assessing how AI algorithms are helping, harming, or hollowing out their work experience and their career journey.

**If any of these three are missing,** you are inserting intentional friction, conflict, and combustion points into your system.

- **Time-bound reviews** for pay, promotion, and reputation impacts.

- **Building and scaling participatory dashboards**: Enabling employees to track metrics they care about (e.g., work-life balance, learning goals) alongside organizational goals. Radical Transparency: Seeing themselves and what they care about in performance tracking tools is foundational to them trusting you and your use of algorithms.

Heather Price, Co-CEO of Symmetra, a global inclusion consultancy, echoed the advice we heard from so many: "The organizations that succeed won't be the fastest to adopt AI, but **those who built trust while implementing it.** Success or failure is really contingent on leadership. The critical question is not what AI can do, but how will leaders choose to use it?"

## TECH STACK, WHAT SUPPORTS THIS

Tooling that tracks AI use/data, logs decisions/anomalies, supports assessments, and routes issues to humans with deadlines — like air traffic control for AI in people systems.

- Model fact sheets: Purpose, data, limitations.
- Bias/fairness testing tools pre- and post-deployment.
- Case management for appeals/corrections, visible to all.

## PHILOSOPHICAL GROUND

Past performance tools have historically been black boxes to workers. Now AI is being integrated throughout their work lives, so it's time to bring everything into the open. Radical transparency is mandatory for trust, and for tracking how power is shifting.

That means tech solutions need to solve real people's problems, not just implement AI for the sake of efficiencies.

That's what Pankaj Rai, Group Chief Data and Analytics Officer at Aditya Birla Group, focuses his team on.

"In my team," shares Rai, "we talk about solving AI problems as a full-stack human being. We empower people to be part of the problem-solving equation by asking them to address Who, What, and Why: Who, defining the target audience we're trying to solve for; What, understanding what they need; and Why, understanding the purpose or reason the solution matters. We solve for the whole problem, and not just the technical problem."

## MONDAY MORNING STARTER KIT FOR C-SUITE, LEADERS

AI exposes the hidden flaws in your current assumptions and evidence-gathering, evidence-making systems. Be fully aware of that, going in... Treat this approach completely differently from anything you've done before.

- Board resolution: No AI-driven negative employment decisions without a human owner and appeal process.

- Convene a cross-functional **"People Scientist Tiger Team"** (HR, L&D, IT, Ops, engineer designers, behavioral science, worker reps) to map your AI landscape, and recommend Stop/Go gates.
- Embrace that, at its core, algorithmic justice is NOT about a dimension of technology: **You are redesigning work**. You are building the foundation for Human+AI partnerships.

**Uncomfortable truth:** Bottom-up justice is crucial to workforce relationships.

## ORGANIZATIONAL KEY METRICS

Are You Delivering Algorithmic Justice?

### Process Metrics

- Percentage of AI systems in live registry (≥90% in 6–12 months).
- Percentage of high-stakes AI evaluated, pre-deployment, as either Stop/Go by Worker Impact Assessments.
- Median time from complaint/appeal to remediation ("Time-to-justice").
- Percentage of AI decisions with named human owner (No fully automated high-stakes decisions).
- Audit frequency/findings fixed (XX%).

### Business Results

- **Increase Return on Future (RoF)** of AI-driven workforce systems by 20–35% within 2–3 years as a live registry, Worker Impact Assessments, and Algorithmic Justice Panels ensure that new AI investments are explainable, auditable, and durable enough to scale across roles, regions, and future use cases without constant rework.
- **Reduce long-tail regulatory, legal, and reputational risk** from AI-managed people decisions by 25–50% over the next 24–36 months, protecting the future option value of AI platforms by embedding contestability, bias testing, and human ownership into every

high-stakes algorithm that touches hiring, pay, promotion, or termination.

- **Improve the success rate and scalability** of next-generation AI use cases by 25–40% in 2–3 years as trusted, justice-centered governance builds workforce confidence, unlocking richer data, better adoption, and faster learning cycles that compound into a structural RoF advantage over slower, compliance-only competitors.
- **Strengthen trust and engagement scores** by 10–15 points where Algorithmic Justice Panels, including worker reps, have real authority to pause, fix, or reshape harmful systems.

## Safeguards Within Justice Team's Metrics

- Team must answer: "What question does this data answer, and how does it improve the system vs. just watch individuals?"
- Open, ongoing bottom-up performance/development conversations with doers.
- Employee contests of results contributed to data accuracy, and actions taken.
- When the Justice team asks, "Does this data help us improve both organizational and individual performance?" ≥80% of the time, the answer should be Yes.
- When leaders ask, "Does this data help us manage results?" ≥80% of the time, the answer should be Yes.

## MONDAY MORNING STARTER KIT FOR INDIVIDUAL CONTRIBUTORS

Ask your leaders: "Can I see what data categories go into our People Analytics?" ...and, "When will you have a policy in place that makes decisions that affect me made by algorithms explainable, and, if need be, contestable?"

**Uncomfortable truth:** If you get evasive replies or continued inaction, plan your exit.

**Individual metric:** The responses you get tie directly to how much you should (or should not) trust your leadership team.

## CLOSING REMINDER

Algorithmic justice isn't about AI: It's about power and trust. Greater openness warrants greater trust.

# Human OS Practice 4

## Building Human+AI Partnerships

### MINDSET, WHY WE MUST CHANGE

Your new Human Capital Operating System turns human-AI collaboration into true co-work: Each side doing what it does best, with less supervision.

AI enables distributed intelligence: It collapses coordination costs, moving more decisions closer to the work. Smart organizations centralize principles, sharing insights via AI. And count AI agents among their staff.

For leaders, this means redesigning roles, work, decision rights, governance architecture, and controls. And **figuring out the optimal Human+AI configurations for value creation.**

For employees, it means you can see clearly which parts of your job are supported by AI, which are uniquely human, and where your growth lies. You'll also have new judgment, decision-making, and scaling accountabilities.

This Practice builds upon, and incorporates, all the work done in the first three practices: reinventing AI Era leadership, redesigning people analytics leveraging taijitu commitments, and ensuring algorithmic justice.

**"The biggest mistakes I'm seeing companies make,"** says Chris Atkins, of The Gap Partnership, and former managing director at Nestlé and Coca-Cola-Nestlé JV, **"is investing huge amounts into AI, but not investing equal amounts into the people who have to use it.** That results in misuse, poor adoption, and catastrophic mistakes because they don't know how to use the power of AI."

Adds Themba Chakela, Head of People, Culture, and Change for e-motion, "Yes, AI is the great amplifier. But the solutions we need sit between people's ears. Freeing that up is the true final frontier."

**True Human+AI partnerships start with three basic and blunt questions:**

- What are our humans uniquely good at?
- What is AI uniquely good at?
- How do we redesign work so both get better, together, in partnership?

"AI is changing several things," says Enrique Rubio, founder of Hacking HR, a global learning community and a trailblazer in disrupting people practices. "First, the nature of work that needs to be done. Second, the way we work and what we need to be skilled at. And finally, the way we interact with others. Going forward, we'll be interacting with AI lots more as teammates."

An example of these ideals: During early 2026, Workday ran a pilot that covered all those areas, giving managers a digital mentor that prepared them for higher-value interactions. It democratizes the kind of coaching expertise that was usually reserved for a select few at the top of the organization, while maintaining meaningful human connections. Workday's pilot found that this approach ensured that leaders were better prepared to lead with empathy and insight.[51]

## SYSTEM RULES AND ARTIFACTS, WHAT MUST EXIST

### Redesigning Roles for AI Partnership

- **Push decision rights downward** with clear escalation rules. Define central decisions vs. team autonomy.
- **Democratize insight via AI.** Provide team-level dashboards on capacity, inclusion, learning, and risk, so frontline leaders and employees can act before problems escalate.
- **Redesign leaders as system stewards.** Expect leaders to tune the Human OS — structures, rhythms, incentives — rather than just "inspire" or create fire-fighting behaviors.
- Train people in **AI literacy**: What AI is good at; where it fails; how to question outputs; and when to override.
- **Create human-in-the-loop checkpoints** for critical workflows.

- **Celebrate human contributions** (judgment, empathy, etc.) in recognition and rewards.
- **Allow local adaptation within global coherence.** Set principles and guardrails centrally (e.g., "We don't surveil individuals, we design for inclusion"), but let teams choose how to implement within those boundaries.

"We want to begin by working on where humans must make decisions, the line between where humans are needed and where AI is best used," shared Ravi Jayasekera, CPO at Hemas Holdings. "I would encourage everyone to use AI like a sparring partner. You bring something of value, then **ask your AI sparring partner to take it further.** And then be able to challenge whatever comes back at you."

This reflects our research. Co-author Vivienne Ming ran a rigorous experiment at UC Berkeley to better understand how humans partnered with AI. Teams were given one hour to make predictions about events, all drawn from Polymarket, a real-money prediction market.

**The teams that faired the best** — only 5% to 10% — **used AI as a sparring partner.** They pushed back, demanded counter-arguments, and interrogated assumptions.

When the AI expressed high confidence, they asked what it was based on. When they had a strong intuition, they asked the AI to make the case against it and learned from their revealed mistakes. What emerged from that back-and-forth wasn't the human's view or the machine's view; it was something neither would have produced alone.

Result: The Sparring Partners were the only group that consistently rivaled the prediction market's accuracy — and on certain types of questions, they outperformed it. These are the types of partnerships you need to develop.

Small business leaders find it easy to leverage their partnership with AI in this way as they have fewer systems to manage. "AI is like a business partner for me," says Eshan Tudawe, an entrepreneur running multiple AI-driven businesses. "It helps me know what questions to ask. And how to access what others know better than me because they have expertise outside of my domain. Everyone everywhere should have an AI partner!"

## MEET YOUR NEW WORKFORCE

The Dubai Future Foundation created the following model for the "post-Turing era, where machine language and intelligence are indistinguishable from human language and intelligence." It clarifies the five dimensions of your new Human+AI workforce.

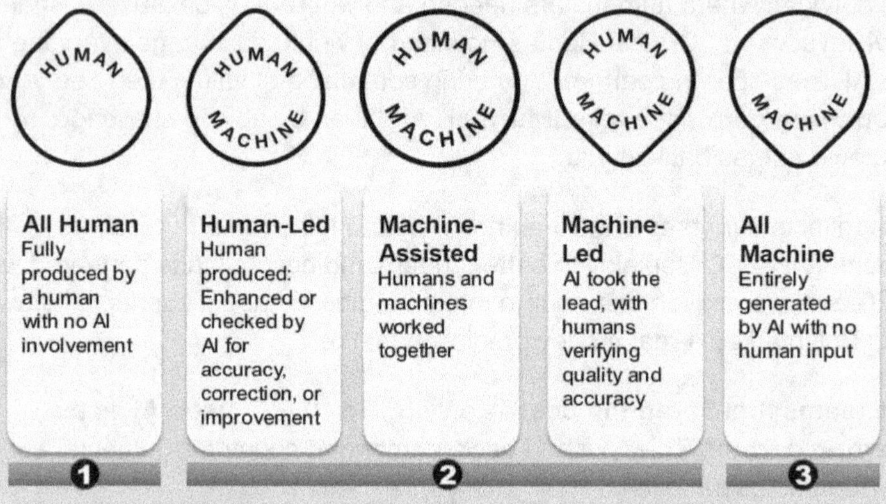


**1.** To determine **Human-Only workflows,** tasks, and metrics... You must first deeply consider, debate, discuss, and answer: "What is the role of our people in the future of work? What are people-only skills?"

**2. For the majority of our To Dos: Human-Led, Machine-Assisted, and Machine-Led work...** You and/or your CHRO plus CIO must completely reimagine all upskilling and development pathways, tools, support, and approaches. Being AI's partner is a new and learned skill; it only comes naturally to a small portion of the workforce.

**3. For all Machine-Only work...** We're confident you'll press hardest on this, and master it as quickly as possible. Because it's mainly about cutting people, cutting costs, and saving money. Just please... Realize it's only one-fifth of the full portfolio of disruptive and transformational Human+AI changes you must make!

## TECH STACK, WHAT SUPPORTS THIS

- **AI-driven system mapping:** Shows decision ripples through the organization. (e.g., "These sales targets are driving burnout without additional support.")
- **Role-based dashboards:** The front-end of all Human+AI partnerships.
- **Rules engine:** Encodes principles into workflows (Human approval, off-limits data, etc.).
- **Lightweight integrations:** AI follows work across silos.

## PHILOSOPHICAL GROUND

**"The future of work will be all about human-AI collaboration,"** one EVP of Data and Analytics told us. "Machines will handle data processing, analysis, repetitive tasks; humans will focus on empathy, connection and collaboration, risk evaluation, and strategic thinking."

This is about creating a true and ultimate partnership, where humans and AI enhance each other to create something neither could achieve without the other.

An example of this grounding is Burger King's testing of its in-headset AI chat bot, Patty. It's designed to help frontline workers improve their connection with customers, "It's really a coaching tool. To help you as an employee become more hospitable...and improve friendliness."[52]

If Burger King is true to their word, fulfilling its promise for better human connections, then this is a great frontline example of Human+AI partnering. However, if they misrepresented Patty's use, and it's also used to invisibly AI-track-and-score people, that violates multiple of the Rights listed in Part 1.

Every step of the AI partnership process, there must be a Values/Principles assessment done by **asking Stop, Start, Continue?** "According to our stated values/principles...

- "To enhance how our teammates partner with AI, what's the **one thing** that we must **stop doing?**
- "The **one thing** we must **start doing?**
- "The **one thing** we must **keep doing?**"

The need for enhanced Human+AI partnering crosses all industries, all organizations, and all customer and employee needs.

Matthew Smith, Customer Strategist for life science firm IQVIA, made that clear in his recap from a recent Pharma USA event: "AI needs to shift focus from efficiency to experience... About 80% of pharma companies see AI as a major revenue driver in 2027 and beyond. However, to date, most have primarily used AI for efficiency and effectiveness gains. Agentic and GenAI have the potential to enable more smooth and seamless customer experiences.

"Pharma needs to be the architects of care experiences, not just innovators in science. The focus should be [using AI] to reduce and remove friction and pain within patient experiences."[53]

**MONDAY MORNING STARTER KIT** **FOR C-SUITE, LEADERS**

**Next 90 days:**

1. **Name the Human Capital OS owner:** List core components, assign executive(s) accountable for coherence. Ensure the owner embraces that the future of work at this company depends on his/her ability to build Human+AI partnership capabilities, enterprise-wide.
2. **Draw Human OS system map:** Cross-functional group maps employee journeys/systems/data flows.
3. **Integration Council:** Monthly reviews, prioritize 2–3 fixes per quarter.
4. **Unified data layer:** CIO/CTO builds People+Work sources.
5. **Gate AI rollouts:** Human OS Impact Assessment required (components touched, behavioral changes needed).

## BUILDING HUMAN+AI PARTNERSHIPS: THREE MUST-DOs

To build partnerships where AI thrives as a copilot teammate to every human's judgment, creativity, brilliance, and relationships, you must...

1. **Do the rigorous pre-work.** Assess for every workstream: **What must be solely human? What must be solely AI?** Design teams must sort through the five possible variations. (See above.) The only way true partnerships will thrive across the organization is for you to do deep human and AI role clarifications ahead of time.

2. **Run Worker Impact Assessments before AI rollouts.** For every high-stakes system, you run a plain-language review of **workload, dignity, bias, privacy, personal development, and job quality**, with worker reps' input on **Stop/Go** gates.

3. **Measure and fix, continuously.** If your Human+AI partnerships fail to deliver desired results, that's not on them. **That's a design, strategy, plan, or role-clarity flaw.** Keep owning that accountability until you get it right.

**If any of these three are missing,** you are failing your partners. Both human and AI.

## ORGANIZATIONAL KEY METRICS

Are You Building Human+AI Partnerships?

### Process Metrics

- Living-systems audits frequency.
- Percentage of AI tools and systems with Worker Impact Assessments.

## Business Results

- **Increase Return on Future (RoF)** of AI and automation investments by 20–35% within 2–3 years as work is redesigned around explicit human-versus-machine task splits, role-based AI copilots, and human-in-the-loop checkpoints, turning point solutions into a scalable portfolio of reusable Human+AI work patterns.
- **Accelerate deployment of new Human+AI use cases** by 25–45% over the next 24–36 months, as teams trained in AI literacy and equipped with role-based dashboards can rapidly prototype, adapt, and standardize Human+AI workflows — cutting dependency on harmonizing teams, and reducing future integration costs.
- **Lift the organization's capacity to pivot** into new AI-enabled revenue streams or service models by 20–30% in 2–3 years, driven by a workforce that is already practiced in Human+AI collaboration, empowered with clear decision rights, and supported by a shared Human+AI operating playbook that can be reapplied to emerging opportunities.
- **Improve decision speed and quality** by 20–30% as frontline Human+AI teams get access to shared, real-time data and clear decision rights closer to the work.

## Human+AI Partnership Team Metrics

- Team must answer: "What question does this data answer to improve how well our humans and AI are partnering together?"
- Quantity of bottom-up conversations and feedback on performance/growth.
- Correlation between contested AI actions and continuous improvement of the Human+AI partnership system.
- When the Partnership team asks, "Does this data help us improve both organizational and individual performance?" ≥80% of the time, the answer should be Yes.
- When leaders ask, "Does this data help us manage results?" ≥80% of the time, the answer should be Yes.

### MONDAY MORNING STARTER KIT FOR INDIVIDUAL CONTRIBUTORS

- **Treat AI as your partner, not your servant** (Insightful and innovative questions in, great insights out)
- **Make technology your apprentice, not your god** (Take control. Make AI your junior partner, subject to your review and hard-won insights.)
- Biggest mistake you can make: Waiting for someone's permission to evolve and grow. (**Creating a partnership with AI is about your transformation,** not just how the company wants you to partner.)

**Uncomfortable truth:** Self-leadership in this partnership means disrupting yourself. Getting comfortable with constantly being a beginner repeatedly.

## CLOSING REMINDER

The secret to great Human+AI partnerships begins with each individual being their best selves, and with what the companies do to support that.

# Human OS Practice 5

## Talent Acquisition Reimagined

### MINDSET, WHY WE MUST CHANGE

**Blow up the old approaches. Set new standards.**

The new future of work journey begins here, in how you seek new people for your teams. Your next breakthrough hire is probably someone your old filters would have missed and bypassed.

**Stop using Lazy AI:** Most current hiring processes just hyper-sped-up your search for keywords in resumes, and searching for “plug-and-play” fits — people who have done that exact role before, just elsewhere.

**Start using Brilliant, Smart AI:** Scouting for raw potential: Searching for the resilient, creative humans who will thrive amid constant AI disruption. Those who have future-proof mindsets, skills, and personal brands. And then combine that with all the non-humans on staff: Humans+AI partnerships, see Practice 4.

**If used wisely, AI widens your lens:** Scanning candidate portfolios, projects, open-source contributions, and unconventional paths that old keyword screens miss, prioritizing learning agility over linear resumes.

This is raising your standards for the AI era, where magical mixes of EQ, resourcefulness, and adaptability create breakthrough value.

Your old systems have acted like bouncers at an exclusive club, checking for narrow career tracks and Day One skills. The Human OS flips to a talent scout seeking humans who will expand your capabilities.

**Opportunity:** AI reveals talent in overlooked places without compromising rigor.

### SYSTEM RULES AND ARTIFACTS, WHAT MUST EXIST

**Redesigning Hiring for Human Potential**

**Phase 1: End of the CV/Resume era.** This job-search artifact was created by Leonardo da Vinci nearly five and a half centuries ago. We’re quickly transitioning away from this relic as a data source, to...

### REIMAGINING TALENT ACQUISITION: THREE MUST-DOs

Are you hiring for yesterday or for the AI Era? If you are to reimagine talent acquisition...

1. **Define hire-for-tomorrow criteria up front.** Every role has a clear mix of **today's skills** and **future-facing superpowers** (learning agility, problem-solving, creativity, collaboration) — and you must hire against both, not just transactional job skills. Most importantly: You are hiring for Human+AI partnering skills, which must be newly built into all searches.

2. **Use AI to greatly expand and upgrade search parameters.** AI is deployed to **find more kinds of talent in more places,** especially in how candidates partnered with AI teammates. Digital deliverables and real-time problem-solving during simulations will be examined.

3. **Measure success differently.** Time-to-fill metrics are no indication of true talent acquisition success in the AI Era. You must measure success by **organizational impact and personal growth at 6—12 months**. Highest-priority hiring process metrics must be by actual impact, not by speed of hiring.

**If any of these three are missing,** you are failing your partners. Both human and AI.

**Phase 2: Expanded digital footprints and records.** The more AI permeates every corner of our lives, Resume Replacements will soon appear. Such as third-party services and apps that scour our complete digital lives to A) certify skills, B) see patterns within our posts, presentations, and work products to identify skills and experience, and C) "connect the dots" between all our experiences and employers' needs for capabilities, personal agility, learning speeds, creativity, imagination, and more.

McKinsey has already embraced this approach during their final-round interviews. Candidates are asked to complete tests using McKinsey's internal AI tool, Lilli.[54] They are expected to prompt the AI, review its output, and apply judgment to produce a clear and structured response. The focus is on Human+AI collaboration and reasoning rather than technical AI expertise.

**Moving from Phase 1 to Phase 2**

- **Define "hire for tomorrow" criteria.** Blend specialist skills with superpower skills (resilience, EQ, creativity, agility, see Practice 8) — weighted for growth potential, not just current fit.
- **Mandate diverse sourcing pipelines.** AI scans beyond LinkedIn: GitHub, community projects, and non-traditional portfolios; require up to 50%+ of candidates from underrepresented paths.
- **Worker-led screening panels.** Frontline reps join interviews to assess cultural fit and real-world potential, with veto contributions on bias red flags.
- **AI as expander, not gatekeeper.** Use for initial sourcing/matching; humans own final calls, with logged rationale and appeal rights.
- **Conduct bias audits at every stage.** AI often just automates biases faster. Pre- and post-deployment bias checks not just on the usual suspects gender, race, age — but also for amazing outliers who might not have exact-match experiences. Auto-flag imbalances >10% for review.
- **Onboarding as a growth launchpad.** Day One includes AI literacy, role mapping (human/AI tasks), and a personal development plan tied to superpower skills.

McKinsey's use of Lilli echoes what many see as seeing as a major shift in talent acquisition. **"AI is increasing the gap between problem-solvers and others,"** says Mihira Kulatunga, head of Marketing for Unilever Food Solutions. "People who are naturally creative, curious, and great problem-solvers benefit greatly from AI. Those who rely heavily on structured processes will struggle. The future of work is changing fundamentally. During the transition, AI will likely increase inequalities between these two groups within your company."

## TECH STACK, WHAT SUPPORTS THIS

- **Broaden your design team.** You must break HR's proprietary hold on this process. Ensure you're including behavioral engineers, I/O psychologists, human-centered design thinkers, and frontline employees.
- **AI sourcing engine:** Scans broad ecosystems (e.g., Behance for creatives, Kaggle for data thinkers); ranks by potential signals like project impact. (Third-party vendors are often used to mitigate privacy concerns.)
- **Skills ontology mapper:** Normalizes software engineer variants; predicts fit via demonstrated outcomes, not keywords.
- **Diversity dashboard:** Real-time pipeline metrics with alerts; simulates offer acceptance rates by demographic.
- **Interview AI prompter:** Suggests behavioral questions tied to superpower skills (Practice 8); records for panel review, not scoring.

## PHILOSOPHICAL GROUND

**The future of work begins here:** How we search for talent sets the standards — and data streams — for what we do with talent once they're inside our companies.

Which means: Leaders must push hard for the complete reimagination and reinvention of talent acquisition. Just bolting AI onto current acquisition practices and processes is future of work malpractice.

This requires a leadership mindset shift, says Raymond Ledwaba, CEO of ITTHYNK. "At the end of the day, short-term, bottom-line needs drive everything else. Unless leaders decide otherwise. **AI is forcing us to think more about the human toll of how we build companies and hire people.**"

**MONDAY MORNING STARTER KIT** **FOR C-SUITE, LEADERS**

**Next 90 days:**

1. **Audit the current pipeline.** Map demographics and outcomes; identify areas for redesigning for human potential, not just role/experience matches.

2. **Pilot AI expander.** Test on one role: source 100 candidates unconventionally; measure quality and diversity vs. your control sample.

3. **Form a hiring council.** Cross-functional leaders plus workers; Train them on potential-spotting; Approve the first redesigned pilot.

4. **Redefine success metrics.** Shift from "Time-to-hire" to "Potential realized at 6-12 months."

5. **Public commitment.** Announce scout mindset shift; track and share progress quarterly.

**Always remember that you're building a pipeline of talent who will create the future.** "At least 60% of the jobs our kids are going to do haven't been invented yet," says former iOCO Group CEO Stephen Van Coller. "The pace of AI-driven change will be incredible."

## ORGANIZATIONAL KEY METRICS

Have You Truly Reinvented Talent Acquisition?

### Process Metrics

- Pipeline diversity (Percentage of non-traditional sources ≥50%).
- Bias audit pass rate (≥95%).
- Worker panel involvement.

### Business Results

- **Increase Return on Future (RoF)** of talent and AI hiring investments by 20–35% within 2–3 years as AI is repurposed from screening for narrow "plug-and-play" profiles to scouting for human+AI potential and future-proof capabilities, building a pipeline of talent that can adapt to roles and business models that do not yet exist.
- **Reduce strategic talent risk in AI-critical roles** by 25–40% over the next 24–36 months by shifting from resume-driven selection to capability portfolios, diverse sourcing, and worker-led

hiring panels — creating a bench of resilient, creative, AI-literate people who can absorb future disruptions without costly external rehiring cycles.

- **Accelerate the organization's ability to spin up new AI-enabled products,** services, or business lines by 25–45% in 2–3 years, as reimagined acquisition pipelines continuously feed in candidates selected for learning agility, curiosity, and collaboration — compounding RoF across every subsequent transformation and growth initiative.
- **Lower manager and leadership churn** by 15–25% as responsibility shifts from firefighting individual burnout to stewarding healthier, regenerative work systems.

### Talent Acquisition Team Metrics

- Team must answer: "Does this data improve system-and-human-potential matching?"
- Amount of bottom-up feedback on process fairness and growth focus.
- Percent of contested decisions creating system improvements (≥70%).
- When the Talent team asks, "Does this data help us hire and retain better partners?" ≥80% of the time, the answer should be Yes.
- When leaders ask, "Does this data deliver continuously improving future capability?" ≥80% of the time, the answer should be Yes.

### MONDAY MORNING STARTER KIT FOR INDIVIDUAL CONTRIBUTORS

**Phase 1, Still in the Resume Era:** Be aware that AI-driven ATS's (Applicant Tracking Systems) absolutely suck! They will relegate most of you to Candidate Zombieland, where fantastic people go to live as eternally undead. They're searching for keywords, and cannot figure out that you're great at something without a very high percentage of exact keyword matches.

For example, one brilliant director-level candidate was rejected because their resume didn't repeat "AI" enough times; said "transformation initiatives," but not "transformation programs;" and spoke about creating value but didn't say "value realization."

For now, always have your AI app optimize your resume for ATS systems, and never expect that a real person will ever see it, until, and only if, you've passed ATS's outdated gatekeeper criteria.

**Phase 2, AI Era:** You or someone you know should be representing the workforce in Talent Acquisition design teams, advocating for great people who might normally be outlier candidates. Self-assess: "How will I fare when my company uses this new process?" Focus your use of, and partnering with, AI tools to improve however you answer that question.

## CLOSING REMINDER

Reimagined hiring builds the diverse, adaptive workforce AI-Era-ready that every company needs.

Completely reimagining is more about leadership courage, vision, drive, and passion, than it is about technology.

# Human OS Practice 6

## Redesign the Workforce's Journey for Humanity

### MINDSET, WHY WE MUST CHANGE

**Well-Being and Personal Agency must be built into system designs, not left to every individual scrambling for themselves.** Every step of the journey, rehumanized.

The people lifecycle at work — hiring, onboarding, performance, development, reward, exit — has, up until now, been built around control, compliance, and cost.

The new Human Capital Operating System rewrites each stage so humans experience agency, dignity, and clarity, supported (not ruled) by AI.

This Practice operationalizes Rights 1 (Agency), 5 (Safety), and 7 (Growth): New-hire interviews and onboarding explore personal goals, workloads include recovery plans, exits offer dignified transitions and ways to improve future journeys for others.

Yes, obviously, corporate needs still matter and remain a priority. But now AI enables individuals' needs — personal goals, health, sustainability — to matter just as much.

AI enables real-time personalization (e.g., onboarding navigators), strain detection, adaptive pathways, and much more.

Says Debbie Goodman, CEO of executive search firm, Jack Hammer Global, "Employees are asking, 'Hold on. I'm training AI, then what happens to me?' AI poses existential questions about why we work. 'Why do you need us humans?'...**'What's my value-added purpose here?'"**

Building on that, one digital transformation leader emphasized that every employee's journey must revolve around their relationships with others. "Human, social connections remain irreplaceable. We need contact. We need touch. We need to belong to a group. That's what makes us human. Even advanced robotics cannot replicate emotional presence, human relationships, mentorship, and trust."

## WORKFORCE JOURNEYS: TIME TO GET REAL

Let's start with the current reality you're building upon. Put yourself in the shoes of your project managers and individual contributors as researched by Gallup, in their State of the Global Workplace 2026 Report.

**Only one-third are thriving. Around one-fifth are pissed-off, sad, and lonely.** Does that bode well for your AI implementation plans?

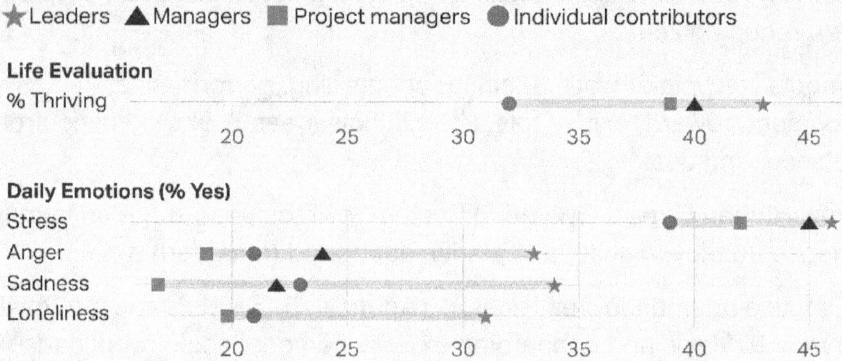


Gallup's *State of the Global Workplace: 2026 Report*

And, oh yeah, **one-third of leaders and pissed-off, sad, and lonely, too!** Any AI plan that's going to succeed has to start by addressing what structural violence and "dying for a paycheck" is doing to the people who will have to implement your plans!

## SYSTEM RULES AND ARTIFACTS, WHAT MUST EXIST

### Human-Centered Journey Milestones

AI isn't just coming for individual lives. It's threatening to dismantle entire career ladders.[55] We must radically rethink how employee journeys support both each individual and the organization in surviving and thriving during this disruption.

- **Hiring:** Assess life goals alongside skills; AI widens diverse sourcing to ensure fit as well as personal agency.

- **Onboarding:** Prioritize belonging and safety; AI guides tool and system designs, not just compliance. First quarter target: Ensuring mastery of company digital spaces as well as a deep focus on community-building connections.
- **Performance:** Deep conversations on superpower skills (Practice 8) and well-being; metrics are balanced (output + sustainability).
- **Development:** Adaptive paths with reskilling; dignified and agency-driven transitions, ongoing support.
- **Workloads:** Built-in recovery and learning times (e.g., 15-20% capacity); AI flags overload for systemic fixes.
- **Expectations:** Define speed and availability clearly; protect downtime, no stealth penalties.

"AI will keep evolving, but the organizations that stay human-centered, where technology amplifies people rather than replacing them, will be the ones that build sustainable performance and trust," says Donald Mokone, director at a multinational equipment repair and maintenance firm. "The real opportunity is operational efficiencies. It's in creating workplaces where people can focus on meaningful, high-value work that requires intentional leadership, not just advanced tools."

## TECH STACK, WHAT SUPPORTS THIS

- **Lifecycle platform:** Integrate stages (HR + collaboration + performance + learning); AI personalizes while eliminating functional silos.

- **Strain detectors:** Calendar and communication analysis for burnout risks (team-level, not individual spying).

- **Journey mappers:** Spotting drop-offs, harms; nudges for balance. Framed as "How can we help?" Providing options for support resources.

## PHILOSOPHICAL GROUND

Workforce journeys reflect your values: Are you leveraging AI to extract the most you can out of everyone, or to create a flourishing and full-agency environment? AI amplifies your intents, values, and process designs.

## REDESIGNING THE WORKFORCE JOURNEY: FIVE MUST-DOs

If you are to redesign workforce journeys for the AI Era, you must...

1. **Incorporate THEIR views of success, not just yours.** For real: What are their dreams, passions, personal priorities, and goals? Every employee journey must be built upon enhanced personal agency — how your company is a vehicle for THEIR success. Enhancing their ability to thrive in a highly disruptive world. Unlike any previous tool or process, AI co-pilots can now help them achieve what's important to THEM, in individualized and personalized ways.

2. **Map real-life, not just HR stages.** Their AI co-pilot can help overlay and incorporate their real-life journey — becoming more resilient, upskilling, caregiving, illnesses, starting a family, buying a home — with organizational milestones such as onboarding, training and development, performance reviews, and promotions.

3. **Intensify design emphasis for transitions.** None of us will ever again be in our Current State for more than a year. All of us will be in a constant state of moving and growing from one state to another. Mastering From/To transitions will be everyone's constant priority.

4. **Leverage AI co-pilots to help individuals explore options.** The hardest part of making any From/To transition is evaluating and making choices. "Do I do A, B, or C?" This will be among your AI co-pilots' biggest value-add.

5. **Protect the sanctity and safety of employee-centric journeys.** You must start designing employee journeys according to taijitu principles (Practice 2), where you place as much effort into how each individual succeeds on THEIR journey, as you do on how they contribute to the company's journey. For personal data security and privacy, you may find that a third-party vendor lifecycle platform is needed.

**"KPIs need to change,"** says Shalini Seneviratne, Marketing Director at Kimberly-Clark. "If AI can produce things faster and more efficiently, and the value we can add is emotional connection and problem-solving, KPIs and how we design jobs to be more fulfilling experiences need to change. We need to start tracking how our curiosity, judgment, and decision-making are leveraged."

Earlier we mentioned Anthropic's study of AI usage, *What 81,000 People Want From AI.* The biggest story may not have been in their findings, but in their methodology. Not your typical static survey. **The AI interviewer** was pre-loaded with hypotheses, but their questions dynamically changed as it learned more from each respondent. Creating qualitative depth and quantitative scale at a fraction of the cost. **This is the future of how you will deeply understand employee journeys and collect their feedback.**

**MONDAY MORNING STARTER KIT** **FOR C-SUITE, LEADERS**

**Next 90 days:**

1. **Co-design one stage** (e.g., onboarding) with recent joiners, managers, and cross-functional leaders: "How can we help ensure agency, belonging, and support?"

2. **Set-healthy-rhythms pilot:** Help people reduce back-to-back scheduling, protect focus and resiliency; measure output + health.

3. **Commit to one difficult change:** e.g., Reduce overload (e.g., project caps); track retention and engagement.

## ORGANIZATIONAL KEY METRICS

Have You Redesigned Employee Journeys?

**Process Metrics**

- Lifecycle satisfaction (e.g., ≥85% honoring each individual's agency).
- Burnout flags resolved within ≤30 days
- Uptake impact created by role and career transition support.

## Business Results

- **Increase Return on Future (RoF)** of human capital and AI investments by 20–35% within 2–3 years as every stage of the workforce journey (hiring, onboarding, performance, development, transition) is redesigned around agency, wellbeing, and AI-enabled personalization — turning each employee lifecycle into a compounding capability asset, rather than a one-off cost.
- **Reduce future replacement and rehiring costs in critical roles** by 25–40% over the next 24–36 months by using AI to design humane, sustainable journeys that improve early-tenure success, deepen attachment, and make internal redeployment and reskilling the default path when roles change, rather than external churn.
- **Accelerate the organization's ability to reconfigure its workforce** for new AI-enabled strategies by 25–45% in 2–3 years, as transparent, human-centered journeys create a workforce that trusts the system, engages with reskilling pathways, and is willing to move across roles, teams, and geographies when the business needs to pivot.
- **Lift engagement and experience scores** by 10–15 points as workers see that their real journeys reflect their goals, realities, and bottom-up feedback.

## Journey Design Team Metrics

- Are AI-driven data aiding and enhancing workforce journeys? (≥80% Yes).
- Work teams: New journey designs and tools support individual flourishing? (≥85% Yes).
- Leaders: Do new journey designs and tools deliver sustainable talent? (≥85% Yes).

### MONDAY MORNING STARTER KIT FOR INDIVIDUAL CONTRIBUTORS

Expect more fulfillment and joy and working smarter, not harder, from your journey!

But you also need to **bring more** to your relationship with work:

- During the reimagined interview process, are you able to tightly articulate your personal dreams, goals, needs, and desires in a work environment? **Most people can't.**
- Are you able to define a long-term (3-5 years) goal about the kind of impact you want to have on others and on the world, through your work? **Most people can't.**

**Uncomfortable Truth:** Yes, you should expect more from companies on your journey. But your responsibilities will also grow in what you need to do to chart your own path.

**What You Gain:** Greater personal agency.

## CLOSING REMINDER

Redesigned employee journeys make AI a partner in human thriving, from start to exit. And AI interviewers will create completely new ways to understand how employees feel about their journeys, and what they need to be their best.

# Human OS Practice 7

## Workforce Planning Reimagined

Build for Adaptability, Not Just Headcounts and Efficiencies

### MINDSET, WHY WE MUST CHANGE

Workforce planning used to be a bunch of number-crunched targets: How many bodies do we need? What will they cost? How do we minimize that cost while maximizing output?

We forecasted headcount. We budgeted for roles. We treated humans as interchangeable units filling predefined boxes on org charts.

Then the world started changing faster than our planning cycles. Roles became obsolete before we could fill them. Skills we hired for yesterday became irrelevant today. The careful headcount models we built in January were fiction by April.

AI has exposed how naïve, inefficient, and wasteful that approach was.

**The New Reality:** In the AI era, the first question isn't "How many people do we need?" It's "What capabilities must we build, and how quickly can our people learn and leverage them?" ... "And after that, how did the world just change, and what do we need to reimagine?"

Headcount is incremental. Capability is dynamic.

**A new way to see and think:** The Human OS mindset shifts to ecosystems: Diverse capabilities, local intelligence, built-in slack for learning and recovery, like a forest that thrives amid storms.

Building capability portfolios future-proofs the organization. And they operationalize multiple workforce-centered Rights: 1 (Agency), 5 (Safety), 7 (Growth), 10 (Value Share).

### SYSTEM RULES AND ARTIFACTS, WHAT MUST EXIST

#### Capability-First Design

- **Retire job descriptions.** Replace them with dynamic skills maps: what outcomes matter, what Human+AI capabilities deliver them.

## REIMAGINING WORKFORCE PLANNING: THREE MUST-DOs

Your reimagined strategy must, at the very least, do this...

1. **Your central unit of planning is now skills, capabilities, and human superpowers.** Gone is the sweating over FTE estimates, and at what cost. (AI can perform that transaction in nanoseconds.) Instead, you're asking, "What capabilities do we need to navigate anticipated and unanticipated disruptions?" Gone are the slots to be filled. Instead, you're asking, "What portfolios of skills have we already built? Which are missing? And how could we build better Human+AI partnerships to meet the anticipated demand?"

2. **You use AI to run real-time, continuous, rolling workforce planning cycles.** That respond to signals from customers, operations, inventories, market conditions, your Human Capital OS, and the workers themselves. Gone are the days of scheduled workforce planning cycles done in executive towers.

3. **Measure success differently.** Outdated workforce planning tracked how quickly you could move great people into necessary positions. AI Era success is how quickly you can move the organization with great Human+AI partnerships — the building of future-ready capacity. "Planning" is now proactive and reactive activation of that capacity, tailored to the types of challenges and disruptions you face.

- **Shift from role-based forecasting to capability-based planning**: Not "How many software engineers do we need?" Instead, "What skills do we need for tomorrow's work? How many pathways exist to build them? What adjacencies enable mobility?" AI models the gaps and builds different scenarios to fill them.
- **Create capability portfolios instead of succession plans.** Traditional succession planning asks "Who is ready to replace whom?" Capability portfolio planning asks "Do we have enough depth in critical capabilities to weather disruption?" Map your capability risks the way you map financial risks.

- **Let the edge think. Make human-AI teams your default unit of work.** Stop pretending HQ knows best. Empower small, cross-functional teams partnered with AI agents to solve real problems close to the customer and the work. Set simple guardrails, give real decision rights, and let adaptability happen at the edges. In the AI Era: **Leaders orchestrate intelligence and decision-making; they don't dictate it.**

### Regenerative Capacity

- Build 15% slack into every team's capacity: Explicitly for learning, recovery, innovation, not emergencies.
- Redefine productivity: Human Energy × AI leverage = Measured by learning velocity and adaptability, not hours logged and paid for.
- Plan transitions transparently: For every AI-disrupted role, define reskilling paths, redeployment options, and support timelines.

### Uncomfortable Truth: Gulf Between Leaders and Workforce

Many CEOs still see AI as an "efficiency accelerator"; many employees see, feel, and fear AI as "replacement accelerator." Reimagined workforce planning bridges those views by making growth and capacity changes visible and shared.

## TECH STACK, WHAT SUPPORTS THIS

All driven by AI, with approaches never before possible...

- **Capability platforms** blend skills ontologies, work data, and market signals for real-time gap analysis.
- **Scenario simulators** modeling "If we automate X, what capabilities rise?" with human override.
- **Talent marketplaces** linking internal skills to needs, tied to reskilling journeys.
- **Regen dashboards** tracking slack usage, burnout risks, and capability velocity.

## PHILOSOPHICAL GROUND

AI forces a choice: Factory (predict, control, extract), or Forest (diversify, adapt, regenerate)?

**MONDAY MORNING STARTER KIT** **FOR C-SUITE, LEADERS**

**Next 90 days:**

1. **Audit one function's planning.** Map current headcount vs. future capabilities; identify 3–5 gaps tied to AI shifts.
2. **Commit to slack.** Pilot 15% protected capacity in 2–3 teams; track output + well-being + learning.
3. **Launch capability conversations.** Quarterly reviews: "What skills are we building? Where are transitions needed?" Publish outcomes.

**Uncomfortable truth:** If your workforce planning still feels like a spreadsheet exercise, you're not ready for AI. You're still pretending stability exists.

## ORGANIZATIONAL KEY METRICS

Have You Redesigned Workforce Planning?

### Process Metrics

- Percentage of AI use cases that include a funded reskilling/redeployment plan for affected workers, not just headcount savings.
- Percentage of roles with skills maps and transition paths (≥80%).
- Internal mobility rate for AI-impacted talent (≥30%).
- Slack/learning capacity utilization (15–20%).

### Business Results

- **Increase Return on Future (RoF)** of workforce and AI spending by 20–35% within 2–3 years by shifting from headcount-based planning to capability-based planning, ensuring that every major automation or AI investment is paired with funded reskilling and redeployment paths that preserve and compound institutional know-how instead of writing it off.

- **Cut the strategic lag between market disruption and workforce readiness** by 25–40% over the next 24–36 months, as dynamic skills maps, internal talent marketplaces, and protected slack for learning enable faster redeployment of people into emerging AI-enabled products, services, and operating models.
- **Reduce future restructuring and redundancy costs** by 25–45% in 2–3 years by treating workforce planning as a living Human+AI capability portfolio —continually reallocating and upskilling people in line with scenario-based AI forecasts, rather than relying on episodic, reactive headcount cuts that destroy future capacity.
- **Increase workforce adaptability** by 30–50% as dynamic skills maps and capability portfolios replace static job descriptions.

## Workforce Planning Team Metrics

- Team able to answer: "Does this approach to planning build organizational and individual resilience?" (≥85% Yes).
- Leaders able to answer: "Does this approach support strategic adaptability?" (≥85% Yes).

## MONDAY MORNING STARTER KIT FOR INDIVIDUAL CONTRIBUTORS

Are you able to name your top three personal capabilities that contribute to organizational success? Have you asked about how to enhance them?

**Uncomfortable truth:** Companies are adopting this approach to win in the marketplace. Own your portfolio of capabilities: Strengths and weaknesses.

## CLOSING REMINDER

Reimagined planning turns disruption into advantage; Humans thriving amid change.

# Human OS Practice 8

## Seven Human Superpowers as Your Strategic Infrastructure

### MINDSET, WHY WE MUST CHANGE

#### Upskilling Everyone: Building a Capability Engine

**1980s to 2020s Employment Bargain:** "Give 110%, and hopefully you get to keep your job. Learning is your responsibility. Figure out what you need to do to stay relevant."

The result: People whose skills became obsolete were made to feel it was their fault for not staying current. Organizations lost capabilities and blamed workers for not adapting fast enough.

#### The New, AI Era Obligation

Jensen Huang, NVIDIA CEO, nailed it at the World Government Summit in 2024: "It is our job to create to create computing technology such that nobody has to program, and that the programming language is human.

"This is the miracle of artificial intelligence. The people who understand how to solve a domain problem — [for example] in digital biology, or in the education of young people, or manufacturing, or farming — who understand domain expertise, can now utilize technology that is readily available to [everyone]. It is vital that we upskill everyone. And the upskilling process, I believe, will be delightful, surprising."[56]

Learning at the speed of AI transformations can't be something people do in their spare time. It must be enabled and enhanced by the system every day, be properly resourced, be consistently rewarded, and be a core infrastructure component.

#### The Seven Human Superpowers

While experts offer different variations, there are seven core uniquely human skills[57] crucial to thriving in an AI-driven future:

- **Self-Awareness:** When disruptions are the daily norm, it's crucial to know what makes you, you — your strengths, fears, passions, super-skills, and under-developed skills. This skill is crucial because it's where you build your intuition — trusting your gut when data is scarce or there are multiple paths you could pursue. Reality check: About 80% of us have only done 20% of the work to achieve valuable and necessary self-awareness.[58]

- **Clear, Concise Communication, Compelling Presentations:** In a world on overload, running on Attention Economy demands, we must all cut through clutter and capture attention. Not by copying-and-pasting AI content, but by making human-to-human connections, and sharing in purely human ways. Your storytelling skills are uniquely human.

- **Creativity, Curiosity, Imagination:** This is the superpower we all had as kids before fitting in was pummeled into us. It remains your superpower in the age of analytics and algorithms, as you are able to combine and connect ideas in ways AI wouldn't think to try.

- **Critical Thinking, Best Questions:** This is your competitive edge in the age of AI. "Some see the world as it is and say Why? I see the world as it could be and say, Why not?" ~ Robert F. Kennedy

- **Personal Agility: Rapidly Unlearn, Relearn:** Hard for many, crucial for all. This enables you to stay useful when the rules change constantly, and allows you to get comfortable always having a "beginner's mind;" leaving you constantly hungry to learn.

- **Empathy:** Keep humanness, connections, caring, kindness, and community in the forefront of everything you do. This is the source of EQ, your emotional intelligence: Reading the room and building relationships.

- **Courage:** Acting despite fear or leaping at the risk of looking foolish. Our shared future needs employees speaking up, speaking out, speaking truth to power, and taking stands for ideas that benefit the company, everyone in it, and its customers.

### Why These Seven Are Foundational

In the AI Era, where tools can outthink us at inhuman speeds, these skills are no longer just owned by individuals. They are organizational capabilities, foundational to your Human Capital Operating System.

AI is rapidly learning to draft documents, crunch numbers, summarize meetings, and optimize schedules. What it cannot do is be self-aware, courageous, deeply empathetic, or morally accountable. The skills that AI can't steal become your core IP.

The companies that are best at hiring for, developing, and leveraging these skills win. It's that simple. And that's hard.

"The things that make us human will become much more important instead of much less important," Anthropic cofounder, Daniela Amodei, recently shared. "And what I mean by that is when we look to hire people at Anthropic today, we look for people who are great communicators, who have excellent EQ and people skills, who are kind and compassionate and curious and want to help other people."[59]

Belgian-based leadership coach, Catherine Herssens, who spent 21 years at Unilever, says, "The most valuable roles will be in things AI can't truly replace: Leadership, creativity, relationship-building, abstract thinking, purpose-driven work. **AI cannot replace your ability to ask the right questions and interpret answers meaningfully.** And you and your story matter. Eight billion people experience the world differently. How you tell and use your story will always be unique."

### Upskilling and Reskilling Must Be Experiential

Our team has seen this firsthand: A global financial services firm launched a well-funded upskilling effort for 1,000 employees through traditional training methods. The program was a catastrophic failure. Within two years, nearly all of the thousand had left the company because they saw the effort for what it was: lateral moves, with a lot more effort just to maintain their same professional status.

Instead, our research — cross-referencing data on 122 million professionals with children's long-term healthy-life outcomes — found a rich set of nearly 50 factors that, collectively, impact our ability to learn. They are meta-learning factors, like emotional intelligence, social skills, and creativity, that shape who you are and who you can become.

## HUMAN SUPERPOWERS AS INFRASTRUCTURE: FIVE MUST-DOs

If you are to leverage the seven superpowers as a competitive advantage, then you must, at the very least, do this...

1. **Name the critical few that drive your uniqueness and competitive position.** Select the superpowers for your best future. (For example: Judgment, courage, collaboration, imagination, moral backbone) and **hard-wire them into roles, workflows, and development.** Pre-AI, these may have been posters on the walls and indoctrinated as cultural norms. Now, not only are they people-to-people norms, but they're also built into algorithms, upskilling, and your Human Capital Operating System.
2. **Strip away work that strangles them.** You deliberately **use AI to remove low-value, high-friction tasks** that block those superpowers — like admin clutter, status reporting, governance check-ins — freeing time and attention for real human work.
3. **Give each superpower a home in your OS.** For every superpower, you can point to at least one core system (planning, performance, rewards, learning, career paths) where it's explicitly supported, measured, and protected.
4. **Fund human growth like you fund technology.** Investment in developing people's superpowers is not a discretionary perk. It is a line item in strategy and capital plans, governed with the same seriousness as your AI and platform spends.
5. **Stress-test your superpowers.** Run simulations, crisis drills, and future-proofing scenarios that force people to use and grow these superpowers in real conditions — making changes and improvements based on the results.

**If you don't** build structures, algorithms, workflows, budgets and Human+AI partnerships around human superpowers, **they are not infrastructure.** They are aspirations waiting to be cut next budget cycle.

Most of these factors involve **experiential learning**. Not instructor-led lessons or e-learning sessions, but **learning by doing** — pushing through failures to achieve success, and to uncover the lessons to be learned ourselves.[60]

This is why developing these superpower skills must be built into how you develop Human+AI partnerships (Practice 4). Supporting and developing how we spar with, and get the most from, our digital teammates is one of the greatest upskilling efforts companies can create.

**SYSTEM RULES AND ARTIFACTS, WHAT MUST EXIST**
**From "It's all on you" to "We're in this together."**

- **Commit to the seven as core OS capabilities.** Make self-awareness, communication, creativity, critical thinking, personal agility, empathy, and courage the backbone of your assessments, curriculum, and development plans — not side electives.
- **Define a clear, shared language** for each skill with simple behaviors and examples at different levels.
- **Embed these skills** into every people process: Hiring, performance, promotion, succession, learning, and leadership development.
- **Co-design learning with employees.** Hold L&D hackathons where cross-functional teams map what great personalized learning journeys look like for different personas.
- **Create personalized learning pathways.** Use AI to map individual skill gaps, identify relevant resources, and suggest development paths based on career goals and organizational needs.
- **Deploy a Personalized Learning Agent** for every employee. An AI coach that can deliver short, relevant learning bites based on current work, career goals, and organizational priorities — like Spotify for growth.
- **Recognize learning as productive work.** Allocate ~15+% of work time explicitly for learning. Protect it in calendars. Measure it in workload planning.

- **Leverage AI to embed learning into work.** Build learning time into project cycles. Create learning sprints between initiatives. Rotate people through stretch assignments.
- **Turn everyday work into practice grounds.** Use micro-prompts in workflows: "What are we missing? ... Who is affected? ... What could go wrong?"

**Cross-Over Integration:**
**Reimagined Learning Drives Reimagined Performance Management**

Traditional performance management created performance theater. People game the metrics that companies look for. The fix: Integrate learning into performance. Move from continuously monitoring performance to quarterly dialogues. Assess contributions over meaningful horizons, not minute-by-minute tracking.

## TECH STACK, WHAT SUPPORTS THIS

Each of the following is a subset component of personalized dashboards:

- Skills maps linking roles, projects, and learning resources
- Recommendation engines matching people to opportunities based on their skills and aspirations
- Career-coach interfaces that surface options and opportunities in plain language
- Learning platforms that recommend targeted experiences for each human skill
- Nudging tools integrated into workflows for reflective questions
- Assessment tools capturing peer observations over time

## PHILOSOPHICAL GROUND

Organizational learning velocity is your AI-era competitive advantage. Organizations that treat learning as infrastructure adapt at the speed of disruptions. Those treating it as an individual's responsibility ossify.

**MONDAY MORNING STARTER KIT** **FOR C-SUITE, LEADERS**

**Next 90 days:**

- Prioritize two of the seven skills for pilot projects.
- Pilot AI-driven learning agent + protected learning time on one team (90 days).
- Calculate learning velocity: Time to build critical capabilities? Percentage of workforce developing skills?
- Redesign performance reviews to explicitly discuss/reward at least two of the seven human superpowers.

**Uncomfortable Truth:** The biggest change of all is **not** creating organizational learning systems. It is a C-Suite mindshift: That learning and upskilling are not costs to be reduced and mitigated, but are foundational to the success of your Human Operating System.

**"This was our 'Holy shit!' moment,"** says Kosala Hewamadduma, head of AI transformation in marketing at Flora Food Group. "We realized that everything we do as marketers can now be done by AI. That means rethinking everything, including how we upskill our entire function."

## ORGANIZATIONAL KEY METRICS

Are You Leveraging the Seven Superpowers?

**Process Metrics**

- How quickly you have redesigned your upskilling and development strategies and structures around the future of work skills, such as the seven human superpowers.
- Speed of embedding those seven skills into processes; percent of personalized pathways; amount of learning time protected.
- How quickly you have created personalized learning pathways for all employees, driven by AI co-pilots.
- Whether your career transition and transformation strategies and structures match the rate of disruption hitting your company from the marketplace, which drives the need for the workforce to evolve and grow at the same rate.

- Percentage of leaders whose own behavior has been explicitly reset and monitored to model healthy boundaries in an AI instrumented environment.

## Business Metrics

- **Increase Return on Future (RoF)** of all AI and transformation investments by 20–35% within 2–3 years as the seven human superpowers (self-awareness, communication, creativity, critical thinking, agility, empathy, courage) become a core capability layer, dramatically improving the quality of problem-framing, AI use-case design and adoption across the portfolio.
- **Cut the time required to build new strategic capabilities** by 25–40% over the next 24–36 months by treating human superpowers as organizational infrastructure — backed by AI-guided, personalized learning agents and protected learning time — so that future shifts in technology, markets, or business models can be absorbed without constant external hiring.
- **Lift innovation throughput and optionality for new AI-enabled products,** services, and business models by 25–45% in 2–3 years, as a larger share of the workforce is explicitly developed and rewarded for curiosity, critical questioning, and courage to challenge assumptions — creating a deep bench of people able to spot and execute the next wave of high-RoF bets.
- **Shorten the time-to-build new strategic capabilities** by 25–40% as learning is treated as infrastructure, embedded into workflows and performance.

## Learning Superpowers Team Metrics

- Team is able to answer: “What question does this data answer to improve human-AI partnerships?” (≥80% able to specify)
- Ongoing bottom-up conversations and feedback lead to continuous improvements in learning systems
- Team is able to answer: “Does this data improve both organizational and individual performance?” (≥80% Yes)
- When leaders ask "Does this data help us manage results?" (≥80% Yes)

**MONDAY MORNING STARTER KIT** **FOR INDIVIDUAL CONTRIBUTORS**

Advice from Kosala Hewamadduma (owner of that Aha moment we just mentioned):

- Be a generalist. Be able to connect ideas from all over
- Be resilient, be able to adapt to constant change
- Build your imagination, creativity, emotional, and spiritual skills
- Build communication skills, how you connect people to ideas
- Live on six-month learn-unlearn-relearn cycles

And from us: Demand learning time as a right, not a privilege.

If your organization won't invest in your capability development **at the same speed and depth as all the disruptions you're experiencing daily,** then they are telling you that they see you as disposable. Plan accordingly.

You have to bring your best too. Of the seven human superpowers: **Assess yourself.** On a scale of 1-10, which three are your biggest strengths? Which three are your biggest weaknesses? Be brutally honest. And vulnerable. Your future depends on it.

## CLOSING REMINDER

Rewrite the organizational code so AI finds humans already centered, protected, and growing. We humans need to decide what AI can and cannot do. Then work on what remains, developing each of our superpowers and personal accountabilities.

# Human OS Practice 9

## The Human Capital Operating System Itself: Integration as Strategy

### MINDSET, WHY WE MUST CHANGE

**The Fragmentation Problem:** Most organizations still relegate people-systems into silos. Hiring doesn't talk to Learning. Performance doesn't always connect to Pay. People Analytics doesn't inform Organizational Strategy. Well-Being is separate from Structure and Work Design.

The result: When it comes to human capital needs, brilliant individual components don't form a coherent system. You optimize each piece while the whole remains dysfunctional.

**The Integration Imperative:** A true Human Capital Operating System isn't a collection of best practices or independent processes. It doesn't launch AI based on "use cases."

It's an integrated architecture where each component reinforces the others, creating emergent properties impossible in fragmented systems. (Made far easier than ever before by leveraging AI.)

When hiring connects to learning pathways, performance assessments inform development, compensation reflects contribution quality, analytics reveal the system's health, and inclusion is designed into every touchpoint — you get something qualitatively different: a regenerative system that makes humans more capable over time.

**This bears repeating:** Designing for humanity does not collapse all "people stuff" into familiar constructs such as HR operating models or digital HR platforms. The Human Capital Operating System reimagines governance, decision rights, data flows, power distribution, accountability loops, and more. **So do not try cramming everything that follows into a pre-AI Era HR function. It won't work.**

## The Meta-System: What Integration-as-Strategy Looks Like in Practice

**Using AI to rethink organization design:** AI covertly does to org charts what seeing Earth from space for the first time did to our worldview: It makes the old borders look arbitrary, and makes us question everything about them.

For about a century and a half, senior executives have used "who reports to whom" and organizational structures to control four things:

- Who gets what information
- Who gets to decide and set goals and measures
- Who directs whom
- What budgets are used to pay for all decisions that are made

AI blows holes in all four. When anyone on a frontline team can see live data about customers, risks, and capacity; when an AI copilot can propose multiple strategic options in nanoseconds; when patterns across the whole company are visible on a shared dashboard — the old idea that decisions must crawl up and down a chain of command stops making sense.

For example, some of your youngest employees, Gen Z, are already covertly automating their reporting, customer outreach, or analysis with AI, without telling their manager, which reallocates decision-rights: Choosing what gets done, in what sequence, with what quality bar, outside any formally approved process.[61]

This shifts power toward who can see the system and act on it, not who sits higher on the chart. Which is the core idea behind the Human Capital Operating System.

You still need senior leaders, but you no longer need every insight and choice to pass through the same narrow executive pipelines.

### SYSTEM RULES AND ARTIFACTS, WHAT MUST EXIST

Translate vision to mandate: Every AI move is designed to create integrated Human Capital OS alignment.

- **Design for coherence across the employee lifecycle.** Every touchpoint — from first contact as a candidate through exit as an alum — should reflect consistent values and reinforce system health and vision.

- **Create feedback loops between system components.** Analytics from performance systems should inform learning priorities. Inclusion data from hiring should reshape sourcing strategies. Wellbeing indicators should trigger workload rebalancing. Pay equity analysis should drive compensation decisions. Each system component should both generate insight and respond to insight from other components. (Through AI-driven dashboards and insight-sharing. Actual human leaders still make the ultimate decisions.)

- **Establish unified governance that sees the whole system.** Create a cross-functional body (Board, C-Suite, CHRO, CIO, workers, ethics experts, techies, I/O psychologists, and more, all aided by AI copilots) that reviews the entire Human Capital Operating System regularly. Not individual programs — the integrated system. Ask: Are the pieces reinforcing each other? Where are the contradictions? What's the emergent system behavior?

- **Use AI as the integration layer, not just individual tools.** Deploy AI that connects data across hiring, performance, learning, compensation, and well-being to reveal system-level patterns invisible when looking at components separately. AI should show you how decisions in one domain ripple through others — how hiring practices affect retention, how performance systems affect learning, how compensation affects inclusion.

- **Measure system properties, not just component metrics.** Track emergent outcomes that only appear when systems integrate: Organizational learning velocity, adaptive capacity, trust levels, innovation rate, collective resilience. These reveal whether your integrated system is regenerative or extractive.

- **Design for continuous evolution, not static perfection.** Build feedback mechanisms that allow the system to learn and adapt. When components conflict, resolve the tension. When new challenges emerge, evolve the system. The goal isn't a perfect system; it's a learning system.

- **Inclusion checkpoint:** Fragmented systems create gaps where marginalized people fall through. Integration means no one slips between the cracks. When hiring connects to development, career breaks don't become career death sentences. When performance connects to learning, different work styles get supported, not penalized.

- **Master Blueprint**: Dynamic visual map showing Practice interlocks (e.g., agency metrics from Practice 3 feed upskilling paths in Practice 7). Update quarterly via AI-collaborative sessions.

- **Integration Protocols**: No deployment without an impact scan across 3+ linked elements. AI auto-flags misalignments.

- **Fusion Dashboard**: Real-time view fusing feedback from voice channels, analytics pulses, and metrics, AI synthesizes insights, surfaces tensions (e.g., burnout rise despite productivity). Teams contest and refine.

**"One of the biggest challenges in building integrated AI-driven systems is trust,"** says Chris Atkins, former MD at Coca-Cola-Nestlé JV. "'Do we trust AI's outcomes and agentic structures? Do we trust its sources, the data?' Right now, the way companies are implementing AI is creating more fear and confusion than trust. Before you start implementing AI systems, you have to do a lot of work on clean sources of data."

## TECH STACK: WHAT SUPPORTS THIS

Orchestration layer unifies stacks: e.g., Workday or PeopleSoft + AI middleware (e.g., Azure AI or ServiceNow). Essentials:

- **Health Monitor**: AI-driven dashboard scoring OS integrity (Green: 90%+ alignment; Red: silos are detected). Drill down to root causes.

- **Seamless Bridges**: Secure APIs linking tools (HRIS to learning platforms to communications), anonymized where needed.

- **Simulation Suite**: "What-if" modeling: test changes across full OS (e.g., new AI tool's ripple onto workload and energy).

- **Self-Heal Loops**: AI proposes fixes (e.g., Auto-adjust quotas if metrics flag overload), with human approvals. No more manual stitching.

- **Safety and Security Risks Addressed**: (True for all ten Practices, stressing here it the point of integration.) Crucial. Both in keeping employee and company data safe, and in ensuring that this tool doesn't go off on its own, with constant guardrails to keep AI in service of humans and the directives it is given.

## INTEGRATION AS STRATEGY: THREE MUST-DOs

To leverage your Human Capital OS as a competitive strategy, you must, at the very least, do this...

1. **Integrate around human rights and capabilities, not just corporate-centeredness.** Integration is no longer just about systems and structures that benefit the organization. Integration decisions now also start with, "How does this protect or erode the ten Rights?" And, "How does this enhance our Human Capital OS and our ability to build new Human+AI partnerships?" You refuse integrations that do not follow taijitu principles. (Practice 2)

2. **Tie executive incentives to OS health.** Anything that truly matters get done this way.

3. **Name a cross-functional OS owner or set of owners, and hold them accountable.** One role to govern integrated performance, no longer separate functional compliance stacks.

## THE PHILOSOPHICAL GROUND

This OS reclaims work's soul. Workforce-centered rights aren't checkboxes; they're the OS kernel.

Integration honors co-creation, with humans + AI as partners. Anthropic's 2026 report warns that deskilling will occur without it;[62] Oliver Wyman notes that Gen Z, the first full-on AI Era generation, craves systems that are fairer.[63] Choose abundance for all, or experience scarcity backlash. Your call shapes not just your company's, but society's future.

## HUMAN CAPITAL OS MATURITY MODEL

- **Stage 1: Fragmented**: The ten Practices are isolated, 50% or greater failure rate.
- **Stage 2: Linked**: Basic connections, manual oversight.
- **Stage 3: Intelligent**: AI-orchestrated, adaptive; 4x ROI uplift. Progress metrics are embedded.

## 3 RULES FOR BEING AN AI-FORWARD COMPANY

1. Be very clear about what's up for debate, and what is not. **Don't hedge on AI. Go all in. Fully commit.**
2. **Define what success looks like.** Become a platform company. Tightly integrated.
3. **Ensure you drive a mental-model shift across the company.** ~ from Jeetu Patel, President, Cisco[64]

## MONDAY MORNING STARTER KIT **FOR C-SUITE, LEADERS**

Block two hours today for drafting your Human Capital OS vision. How will you reimagine the future of work for the AI Era? Then...

1. **Name the OS and its owner.** Explicitly define: "This is our Human Capital Operating System." List the core components (hiring, learning, performance, workflows, pay, analytics, well-being, inclusion, governance, decision rights, data flows, power distribution, workforce planning, etc.). Assign a single set of executives as owners (often the CHRO in partnership with CIO/CTO/COO) who are accountable for **system coherence**.

2. **Draw the first Human OS System Map:** Gather a cross-functional group (HR, Finance, Ops, Tech, Legal, key business units, and worker reps). Map the full employee journey end-to-end. Then, list your major people-related systems and platforms. Then connect: "Where does data flow? Where does it dead-end? Where do decisions hand off badly?"

3. **Create an Integration Council with decision rights.**

4. **Consolidate around a unified people-and-work data layer.**

5. **Pilot:** Pick two high-signal areas where contradictions are obvious (for example: performance + learning, or pay + inclusion). Do one visible integration move in each:
   • e.g., Make progression dependent on both results **and** contribution to learning/mentoring.
   • e.g., Have your pay review explicitly reference your diversity and equity commitments, with data on how actual decisions moved you toward or away from them.

## ORGANIZATIONAL KEY METRICS

Are You Succeeding with Your Human Capital OS?

### Process Metrics

- Percentage of the ten Practices that are interlocked and are in active pilots (target: 75% in Year One).
- Integration audit pass rate (quarterly; <80% triggers review).
- Time to resolve OS-flagged tensions (target: <4 weeks).

### Business Results

- **Increase Return on Future (RoF)** of AI and people investments by 25–40% within 2–3 years as fragmented HR, analytics, learning, performance, wellbeing, and AI-governance tools are integrated into a single Human Capital Operating System, turning isolated pilots into a reusable, enterprise-wide platform for future transformations.

- **Reduce future transformation and AI program failure rates** by 30–50% over the next 24–36 months by requiring every major initiative to pass through an integrated OS impact check — surfacing cross-system conflicts (hiring, workload, incentives, data flows) before launch, and protecting the option value of subsequent AI bets.
- **Accelerate time-to-value for new AI-enabled strategies,** products, and operating models by 25–45% in 2–3 years, as unified data, governance, and "fusion dashboards" allow leaders to see system-wide effects in real time and coordinate fast, organization-level adjustments instead of slow, silo-by-silo fixes.
- **Finally addressing zombie automation,** referring to legacy automated systems, software and AI scripts and agents that continue running, creating To Dos for so many teammates, long after they have lost their legitimacy. These 'undead' processes often persist in enterprise systems, creating operational inefficiencies and security vulnerabilities. These zombies are a constant drag on Human+AI partnerships.
- **Water-cooler check:** Hearing lots more conversations about how much easier it is to work smarter and how that's impacting project speeds and results, employee satisfaction, customer satisfaction, sales and revenues

### Human OS Integration Council Metrics

- OS Fluency Pulse: Anonymous score: "Do my tools and skills and rights flow seamlessly within my work?" (track quarterly).
- Personal contributions to fusions logged and recognized.
- Integration efforts having a measurable impact on top-line and bottom-line results.

## CLOSING REMINDER

The hard work behind integration isn't optional; it's the OS heartbeat: coding humanity into everything you do.

Forge it and unleash potential. Ignore it and invite collapse.

# Human OS Practice 10

## Sustainability Reimagined: Your Human OS's Impact on The World

### MINDSET, WHY WE MUST CHANGE

Decent work, gender equality, reduced inequalities, quality education, health, and strong institutions: These Sustainable Development Goals (SDGs) are not abstract global dreams.

They rise or fall in the everyday decisions your people systems make: Who gets hired, who gets trained, who burns out, who gets heard, who gets ignored or pushed aside. AI magnifies these decisions, for better AND worse.

For boards, this reframes SDGs from glossy reporting to operating reality. Your Human Capital System is one of the most powerful levers you have on the SDGs. For employees, it answers a simple question: "Does working here help build a more just and better world?"

**The Disconnection Illusion:** Most organizations treat Sustainable Development Goals, such as the 17 goals created by the UN[65], as external commitments — things you do "out there" through corporate social responsibility, philanthropy, or sustainability initiatives.

Meanwhile, your internal people systems — how you hire, develop, pay, and treat people — are assumed to be separate. Internal operations, "in here" stuff.

**The hidden truth:** Your human capital operating system is one of the most powerful levers for SDG progress that you control. How you design work, distribute opportunity, and exercise power directly impacts:

- **UN SDG 3 (Health and Well-being):** Are your systems regenerative or extractive?
- **UN SDG 4 (Quality Education):** Are you building capability or keeping people around until the expiration date of their usefulness?

- **UN SDG 5 (Gender Equality):** Are you closing pay gaps or perpetuating them?
- **UN SDG 8 (Decent Work):** Are you creating dignified work or grinding people down?
- **UN SDG 10 (Reduced Inequalities):** Are you growing personal agency for all employees, or replicating "out there" economic chasms?

The future of sustainability will belong to organizations that are willing to redesign the invisible architecture of work, and to use technology to lift the futures of all who work within those structures.

**Understand AI's impact on really big issues...** "The survival of humanity depends on AI," says Jeetu Patel, Cisco's president. "Birthrates are going down, and we have a demographic shift where there will be more people in the older age bracket than the younger bracket. At some point you have 60% of your population where you don't have enough people to take care of them. And that could cause a lot of human suffering. We have to be careful that AI's capacity is working on behalf of humans."[66]

### Addressing AI's Horrific Impact

AI may be a savior on some issues, but **sustainability-wise, it's also a destroyer.** AI's high power demands from training and inference phases can drive massive increases in carbon emissions, strain electricity grids reliant on fossil fuels, deplete water resources for cooling, and exacerbate energy affordability crises for households. (SDGs 6, 7, 12, 13, 14, 15).

**There's also AI's exacerbating impact on pre-existing class struggles.** "The tech titans who are building and promoting AI are not worrying about who falls behind," says Adriaan Kruger, cofounder and CEO of nuvoteQ.io. "It seems we're heading to a future where the ultra-rich will just become richer and we'll have a very large lower-middle class. That worries me." (SDGs 1, 2, 8, 10, 11, 17)

*Rolling Stone* nailed that disparity perfectly with their early 2026 article, "The Rise of the Digital Oligarchy: AI has pushed society into tumult and confusion. What comes next will determine if humanity remains the author of its own story, or if it is quietly written out."[67]

If your company benefits from the use of AI, **you have a responsibility to work with local communities and with governments** to mitigate its environmental impacts as well as societal and economic impacts.

There's also the **Digital Matilda Effect,** where women's intellectual contributions are attributed to their male colleagues. A recent study by Harvard Business School found that AI is likely contributing to new gender inequalities.[68]

**Finally, pay attention to social media's legal woes:** As our first edition went to press, some were calling it social media's "Big Tobacco moment." Meta and Google lost a major lawsuit on claims that their products are addictive in the same way that brought down Big Tobacco:[69] Concealing damaging internal research, selling a product engineered to be addictive without disclosing its risks, and aggressively marketing to underage users.

Why should this matter to how you build your company? Because it's all about behavioral engineering built into algorithms. The same path AI inside companies is now on.

## SYSTEM RULES AND ARTIFACTS, WHAT MUST EXIST

Hardwire Sustainable Development Goals into your Human Capital OS core with explicit links.

- **SDG-OS Map**: Blueprint tying the ten Practices to company goals (e.g., Practice 8 upskilling connects to SDG 4). AI updates dynamically.
- **Impact Mandates**: Pre-AI rollout, model SDG ripples (How will AI impact equity gaps? Job dignity?). Pause and rethink if the risks are high, or if you find you're not doing enough.
- **Equity Protocols**: Global baselines: Are pay and learning access uniform across the company? Are frontline employees marginalized or prioritized?
- **Accountability Ties**: Exec pay tied to SDG metric improvements (Gap closures, Improved work scores).
- **Transparency Reports**: Internal dashboards: e.g., "OS moved SDG 10 by X%." Worker feedback on harms is valued and leveraged.

## REIMAGINING SUSTAINABILITY: THREE MUST-DOs

To claim sustainability through your Human Capital Operating System, you must embed at least these...

1. **Make the AI Era ten Rights your longevity guardrails.** They protect long-term human capital priorities even when short-term capitalism screams "cut." We need to protect human sustainability, for all our sakes.

2. **Report human sustainability like financials.** Personal agency, mental health, value-sharing, learning, resilience, and more must sit alongside revenue margins and ESG metrics, with executive comp tied to long-term human capital gains.

3. **Extend your OS beyond your walls.** Audit and influence vendors, gig platforms, and supply chains for Rights alignment, refusing to profit from exploitation you'd never allow internally. Raise the bar for Human Capital OS standards.

### TECH STACK: WHAT SUPPORTS THIS

AI bridges internal and external:

- **Impact Modelers**: Simulate OS changes on SDGs (Does automation drive inequalities, especially where automation reduced headcount?)

- **Unified Dashboards**: HR + ESG data fused: track pay gaps, learning access

- **Predictive Alerts**: Flag risks (if burnout spikes, that indicates SDG 3 failing).

- **Global Equity Tools**: Bias audits across geographies; create anonymized input hubs.

## THE PHILOSOPHICAL GROUND

OS choices ripple across the planet. Dignified work templates impact the future of society. Enlightened Human Capital OS designs make a difference internally and externally.

**MONDAY MORNING STARTER KIT** **FOR C-SUITE, LEADERS**

Treat Sustainable Development Goals as a system integration choice. Over the next 90 days:

1. **Draw your SDG–Human OS map.**
2. **Name your "non-negotiables** for respectful, dignified, full-agency work," newly defined the AI era.
3. **Insert at least one human capital SDG metric** into the core business review.
4. **Tie a slice of executive variable pay** to Human OS SDG outcomes.
5. **Run "impact before efficiency" reviews for major AI decisions.** For any significant automation or AI roll-out (especially those promising big cost savings), require a one-page "Decent Work & Equity Impact" report that answers:
   • Who will see their work changed or displaced?
   • What reskilling or redeployment options are on the table?
   • How might this decision widen or narrow existing inequalities?
   • What is the energy/computing footprint of this system, and can we mitigate it (e.g., through scheduling, infrastructure choices)?
   **If the answer to any of the above is, "We don't know,"** you're not ready for AI implementation.

## ORGANIZATIONAL KEY METRICS

Have You Reimagined Sustainability?

### Process Metrics

- SDG-OS alignment score (target: 85%).
- Impact models are run pre-pilot for any new AI rollouts (100%).

## Business Metrics

- **Increase Return on Future (RoF)** of AI and human-capital investments by 20–35% within 2–3 years as SDG-linked metrics (decent work, equity, learning access, wellbeing) are wired into the Human Capital Operating System and executive scorecards, ensuring that every major AI deployment builds long-term social license, talent attraction, and stakeholder trust rather than eroding it.
- **Reduce medium-term ESG, regulatory, and reputational risk costs** by 25–40% over the next 24–36 months by requiring "Decent Work & Equity" impact reviews for all significant AI and automation decisions — protecting the future option value of AI platforms and business models in increasingly regulated and values-driven markets.
- **Strengthen the organization's access to capital, customers, and talent over** the next 2–3 years, reflected in higher RoF, by demonstrably aligning the Human Capital Operating System with priority SDGs (4, 5, 8, 10), differentiating the company as a trusted, future-fit employer and partner in markets where AI-driven social harm is becoming a major selection and valuation filter.
- **Improve retention and engagement of priority** employee groups by 15–25% as SDG-linked metrics (dignified work, growth, wellbeing) become visible, managed targets.
- **Enhancing your public/private partnerships.** You are moving at the speed of AI, while legislation in local, state, and federal governments moves at a snail's pace. Your employees need you to guide, support, and lobby for AI Era updated human rights — moving your enterprise design and moral principles into legal constructs that advance human rights everywhere.

## ESG/SDG Design Team Metrics

- OS-SDG Pulse: "Is our work advancing the company's values and justice for those who are impacted by our company?" (80% Yes).

## CLOSING REMINDER

Sustainability is not only about what your company does to the planet. It is about what your company does to people — you, and everyone else — every single day.

And your human capital OS impacts your SDGs every day.

# If you remember one thing about these Practices, let it be this...

**If you zoom out, the ten Practices are really one promise about the coming powershift, told ten different ways.**

We will design work
so that technology amplifies
humans,
not erases them.

We are now designing the
full expanse and amazingness
of human experience
into how we work.

And we built the
Human Capital Operating System
to make that happen.

**Here's what that looks like in everyday terms, and in leadership accountabilities...**

- You have 5 to 8 years, maximum, to build completely new AI-native enterprise operating models.

- To get there, put on your non-linear, exponentially-disruptive-change-within-2-to-3-years thinking caps. None of this happens without massive mindset shifts.

- During that time, you will be releasing the most powerful, inscrutable, uncontrollable technology ever invented. You've got to spend a lot of time and effort understanding, and adjusting for, unintended consequences.

- Leaders are gardeners, not bosses with bigger dashboards.

- They steward energy, capability, and trust — and they are judged on those outcomes, not just quarterly financial results. They are building a living system that must evolve with and by AI capabilities. THAT is what future-proof systems are.

- Data is used to understand and repair and enhance systems, not to stalk individuals.

- Dashboards become mirrors everyone can see, not microscopes pointed down from the top.

- Algorithms get human bosses, brakes, and audit trails. Ban "the computer decided." Every important AI-shaped decision has a name and a human story behind it.

- Humans and AI become co-workers, sparring partners. Getting there will take a lot of upskilling, coaching, nurturing, and support. (More than you think good, dependable, hard-working people should need.)

- Machines eat the drudge work; humans do the judgment, empathy, creativity, and courageous work.

- Hiring stops being a keyword-driven and becomes an AI-driven searchlight; hunting for real skills and potential in places your past, too-tightly defined filters never saw.

- The employee journey is redesigned as a human journey. From first interview to final exit, people experience agency, growth, and care — not just compliance and control.

- Stop all (unintended) structural violence. Period. Hold every designer of every system accountable to this new standard. No more "dying for a paycheck."

- Workforce planning shifts from "How many heads?" to "What capabilities, and how fast can we build them?" ... "Downtime" for learning and recovery stops being treated as waste and starts being treated as strategy.

- Learning becomes infrastructure, not a side hobby. The seven human superpowers — self-awareness, communication, creativity, critical thinking, agility, empathy, and courage — are built into the system, not left to chance.

- All the pieces talk to each other. Hiring informs learning; learning informs performance; performance informs pay; analytics reveal health; inclusion is baked in, not patched on.

- And all of it is tied to something bigger than the company. Your Human OS becomes an engine for decent work, equity, education, health, and trust in the wider world.

That is what a Human Capital Operating System can be in the AI Era.

Not a new HR platform. A new way of deciding what humans are for — and what technology is for.

Every principled stand for the ten Practices creates a new, amazing future for everyone who comes after you.

**The roadmap is in your hands. The mirror is already on.**

**What you build next is your answer.**

# Post Script

## What If We Finally Get It Right?

What if everything AI turns out great?

**What will you do with your newfound freedoms, all your full agency, and more time and energy available to make a real difference?**

Most of us haven't thought that far out. We're too busy running faster and faster on the hamster wheel.

Or, as Antonio Melonio, the "anarcho" author from the Balkans, called it... "Living the institutionalized bargain that says: Give us your waking hours, your attention, your nervous system. We will give you conditional access to food, shelter, and the right to not die in the street."[70]

For far too many, that bargain sucks. In multiple studies across the globe, nearly four in ten people show up each day, doing work they think is pointless, and makes no meaningful contribution to the world.[71]

Four in ten. Grinding it out. Feeling it's all pointless.

**AI holds the possibility to change that!**

By changing how we work and why we work. By taking over work that feels meaningless.

Let's close this book by hoping that AI fulfills at least some of that promise. Where some or lots of the pointless, meaningless work is taken over by machines.

**Are you ready?**

**What would you like the new world of work to do for you?**

**What will change for you?**

As Melonio shared, "We have spent ten thousand years building for this moment. The ancient Greek word for the good life was not success or achievement. It was eudaimonia. Flowering. The full bloom of a human being who has the time, enough safety, and enough freedom to become whoever they actually are."

What if humans were no longer labeled "resources," and we actually built companies and worked to deliver eudaimonia?

**No matter where AI goes from here,** let's envision that as what's next. Deprogramming from nearly three centuries of industrial optimization. Becoming more of who we actually are. Let's prepare for that.

We have one final question for you...

**What do you want your legacy to be?**

What will your grandchildren say about the AI-driven future that you helped create, and then left for them?

Decades from now, how we'll be viewed by future generations is not what we did with AI, but what we did to the people it touched.

**Exercise:** Just like companies that — even while going through massive AI transformations — have strategic plans, your personal long-term strategic plan is your legacy: The impact you wish to have on others and on the world.

Three years is a perfect "long-term" timeframe: It allows you to envision the impact you wish to have, and challenges you to act for over 1,000 days, with that goal in mind, not putting it off.

- **To Do: In just one sentence or a short phrase:** What's the impact you wish to have on people and your world three years from now?

- **Write that down.** (e.g., "I want to help every individual and leader I touch be their best and most amazing selves in how they use AI.")

- **Workshop it with friends and family.** Does it feel like you? And does it honor your passions? Does it inspire and motivate you? Does it make the world better?

- **Be crystal clear** on your desired three-year impact.

- **Include** what must be restrained today, or approached completely differently, for you to achieve that three-year goal.

And one final metric...

Who will thank you for the difference that your legacy made in their lives? Who will have more freedom, full agency, and time to be happy because of you?

Now go make that happen! With Ten Rights and Ten Practices.

# Acknowledgments

## Thank you!

**Malisha thanks...** My deep gratitude to my co-authors who helped turn thought into movement. Together, we chose to believe that work can be reimagined, with courage in our hands and care at our core. And to my family, you are my primary source code. Everything I build, every step I take forward, is rooted in love for you.

**Bill thanks...** My amazing family, especially Mom who shaped so much of who I am, and my son, Ian; my legacy. Our wonderful community and my fantastic co-authors. Seven decades of life experiences, uplifted by coaches, mentors, teammates, clients, and friends; all of whom believed in me, sometimes more than I did. And, of course, Star Trek.

**Vivienne thanks...** "Being different makes the world better." I've spent years building the evidence for this, but my daughter and son made me believe it.

**Anish thanks...** I am deeply grateful to my co-authors for their sharp thinking, trust, and shared conviction in bringing this work to life. To the HARE People collective, this book carries your questions, your challenges, and your belief that work can be reimagined with both courage and care. To the colleagues, mentors, bosses, and team members I've had the privilege to work alongside over the years; you have shaped my thinking, challenged my assumptions, and made the work real, not theoretical. And finally, to my wife, Preethi, and my boys, Aadi and Ayan, thank you for your constant support, patience, and the joy that keeps me grounded through it all. This would not exist without you.

**Nadeesha thanks...** I am deeply grateful to the friends and associates I reached out to for their perspectives on AI and the future of work. Despite their busy lives, they generously shared their insights, many of which have shaped this book. What stood out most was a shared sense of purpose: A commitment to ensuring that humanity is preserved, safeguarded, and strengthened through the choices we make today. My sincere thanks to my co-authors for the trust, openness, and shared conviction in bringing this work to life. It has been a privilege to learn alongside each of you and to contribute to something larger than any one of us. To the HARE collective, this book carries your thinking, your questions, and your courage to challenge how work is evolving. Your influence runs deeper than words on these pages. And to my family — my husband, parents, and children — you are my grounding and my strength. Your love, patience, and quiet belief make everything I do possible.

**And we all thank...**

**Our book partners:** Huge thanks to Simon Akerson, Lead Project Director; Sarah Joseph, Project Associate; Sean Hernandez, Book Cover Designer; Martin Garcia, Infographics Designer; Peter Spencer, Design Formatter; and Daniel Gonzalez, Video Choreographer. Despite challenging circumstances, including severe weather conditions and disruptions in Chicago during portions of the production schedule, the team continued to push forward, ensuring the project remained on track.

**Our interviewees and early readers:** Scores of each. To those who shared their stories, even the difficult ones, your truths made the work whole. Thank you! To all who took the time to make sense of our ramblings and offer specific improvements, huge thanks to you, too!

**You!** To the next generation, you stand as proof that technology must lift, not replace: Thank you for inspiring us to fight for you! To the leaders and managers and workers who will take up this cause, making the future of work better, safer, more joyful for all us humans: Thank you! And to all you readers who found that one nugget in this book that ignites something inside of you: This book is nothing without your Ahas, the conversations that you will drive, and the actions you will take based on what you read. Thank you for adding that missing ingredient!

# Endnotes

Human Capital Operating System™ is trademark pending. **Please cite this book and/or its authors when using this term.** Trademarked uses apply to Class 35, organizational design, business transformation, human capital strategy, and workforce optimization; and Class 42, AI workflow integration, algorithm-driven operational frameworks, and tech-enabled systems.

---

[1] McKinsey AI agents as teammates: https://bit.ly/4tO0nzR

[2] Millions of jobs to be displaced by AI: https://bit.ly/3Oc95ld

[3] AI hiring humans: https://rentahuman.ai

[4] Denmark deep fake protection: https://bit.ly/4tJjHyh

[5] Surveillance capitalism: https://bit.ly/4rEBd5d

[6] Amazon "time off task": https://bit.ly/4rp3i0h

[7] Eightfold AI class action suit: https://bit.ly/4b8OYTX

[8] GDPR's right to explanation: https://bit.ly/4c7s9B4

[9] 60% of jobs driven by algorithms: https://bit.ly/3OkTciv

[10] Microsoft productivity scores: https://bit.ly/4qPwMTQ

[11] Carole Cadwalladr, data rights are human rights: https://bit.ly/4kKj0Ay

[12] GDPR's right to be forgotten: https://bit.ly/3MIKolN

[13] AI doesn't reduce workloads, it intensifies them: https://bit.ly/4rCaYws

[14] "We can hack human beings": https://bit.ly/3OAMbdh

[15] Structural violence: Term coined by Norwegian sociologist Johan Galtung in 1969

[16] EU AI Act: https://bit.ly/4qKK43V

[17] AI will disrupt hundreds of millions of jobs: https://bit.ly/3MRqncZ

[18] Block cut nearly half its workforce: https://bit.ly/4aZZJGY

[19] Long work hours linked to heart disease and strokes: https://bit.ly/3MHiQNG

[20] Employee burnout costs global economy $322 billion annually: https://bit.ly/46kmwvC

[21] Nearly eight in ten employees are disengaged: https://bit.ly/4rwlBRi

[22] Algorithmic management without worker is unjust: https://bit.ly/3MTCKoU

[23] AI implementation failure rates, up to 95%: https://bit.ly/3MRtnpL

[24] AI ROIs should include employee's sharing in profits: https://bit.ly/4atimnA

[25] Study advocating four-day workweeks: https://bit.ly/4ufuPTK

[26] AI productivity dividends: https://bit.ly/47rbTaX

[27] $1k per month AI dividends: https://bit.ly/4c9AgMO

[28] Universal basic income: https://bit.ly/4slKbyu

[29] Anthropic $1.5 billion settlement: https://bit.ly/3ZKYMgu

[30] 5-8 years estimate for AI-native tipping point, Near-term pivot: https://bit.ly/4slVF55

[31] Crossing the Rubicon: Historical moment (Julius Caesar, 49 BC) that has come to symbolize a point of no return, when the time for deliberation is over, and difficult and decisive action is required

[32] 5-8 years estimate for AI-native tipping point, Crossing the Rubicon: https://bit.ly/4sEVess

[33] MIT study: AI will handle most text tasks by 2029: https://bit.ly/3OjtXxa

[34] Gartner study: Shift to agentic AI platforms: https://bit.ly/4tsraAL

[35] Frontier AI, McKinsey: https://bit.ly/47uX7Qg

[36] 5-8 years estimate for AI-native tipping point, Likely tipping point: https://bit.ly/3PkFJHH, https://bit.ly/4t58FIV

[37] 5-8 years estimate for AI-native tipping point: Leadership AI investments, views, and strategies: https://bit.ly/4lNjCWR

[38] Algoworks report 2026, Real-world AI adoption and ROI: https://bit.ly/40Dph86

[39] Artificial General Intelligence by 2030s: https://bit.ly/4sNnGZm

[40] One Industrial Revolution mass-use anchor, 1870s railroad construction and use: https://bit.ly/4rs7QlF

[41] Digital Era mass-use anchors: https://bit.ly/4lqv762

[42] GenAI and Agent mass-use anchor: https://bit.ly/4saLlxz

[43] Coca-Cola and Walmart CEOs handing off AI transformations: https://bit.ly/4m3Fuxe

[44] The trillion-dollar blind spot: https://bit.ly/4myO4UW

[45] Companies laying off employees because of AI's potential: https://bit.ly/4tPtFhA

[46] Klarna lays off, then has to rehire: https://bit.ly/4tJJZjX

[47] Report, The World's Most Important Problem: https://bit.ly/4qKXMUJ

[48] Tokenmaxxing: https://bit.ly/4d44mm3

[49] Solution, more women leaders: https://bit.ly/47Tv4uh

[50] Uber employees create a clone of their CEO: https://bit.ly/3MPqSnU

[51] Workday's digital executive coach: https://bit.ly/4dpM4M0

[52] Burger King headset AI chat bots: https://bit.ly/4l6ulpp

[53] Matthew Smith, IQVIA, on pharma leveraging AI to enhance patient experiences: https://bit.ly/4ddqDOr

[54] McKinsey final round of interviews, using Lilli, their AI tool: https://bit.ly/4cGYNtp

[55] AI dismantles career ladders: https://bit.ly/4bUIRla

[56] NVIDIA CEO on upskilling everyone: https://bit.ly/3MzIuEd

[57] Seven superpowers researched by the Jensen Group, ongoing study, The Search for a Simpler Way

[58] 80/20 Rule of Inner Knowingness researched by the Jensen Group, The Search for a Simpler Way

[59] The things that make us human will be more important than ever: https://bit.ly/4aMoPsQ

[60] Vivienne Ming's research and findings as Chief Scientist at Gild and at Socos: https://bit.ly/4uH0UEa

[61] Gen Z creating hierarchy workarounds: https://bit.ly/4aPnr8x

[62] Anthropic 2026 Report: https://bit.ly/4cHaMam

[63] Gen Z craves fair systems: https://bit.ly/3OrTjsx

[64] Podcast, Cisco President Jeetu Patel: https://bit.ly/46sc59i

[65] UN's 17 Sustainable Development Goals: https://sdgs.un.org/goals

[66] Podcast, Cisco President Jeetu Patel: https://bit.ly/46sc59i

[67] The Rise of the Digital Oligarchy and society's growing digital divide: https://bit.ly/4sDMNhj

[68] The Digital Matilda Effect, increasing gender inequalities: https://bit.ly/4vd3a6n

[69] Social media's Big Tobacco moment: https://bit.ly/4sO0Gtx

[70] Living the institutionalized bargain: https://bit.ly/4c8GPQo

[71] Four in ten think their work is pointless: https://bit.ly/3MYQZJ3

# About Your Authors

| **Malisha** Awumor | **Bill** Jensen | **Vivienne** Ming | **Anish** Padinjaroote | **Nadeesha** Wijesingha |
|---|---|---|---|---|

**Humanity** did not rise from one voice; it was shaped in chorus, through the minds and passions of the HARE Community. (**H**eart of **A**utonomous **R**egenerative **E**mergent People) Established by Anish Padinjaroote, focused on designing the future of work, humans first. Find our community at www.harepeople.com.

Between us, your authors have nearly 150 years of experience in cognitive science and behavioral economics research, driving organizational and digital transformations, coaching and developing leaders, and designing the future of work — hard-won, practical wisdom, all focused on human-centered change.

**Malisha Awunor** is a technology-for-good strategist, with 25+ years of global experience across strategy, behavioral science, and the future of work. She is a firm believer that access to technology is not a privilege, but a right. Her work sits at the intersection of emerging technology and humanity exploring how we ensure that, as AI advances, humanity is not diminished but is elevated. She advises boards and executive teams on navigating the shifts AI is bringing to organisational design, digital equity, and the meaning of human value at work. Malisha is the founder of Ikigai People Advisory Services, a director of the FutureShift and Movement-for-Good NPCs, and the creator of the I AM virtual reality bias protocol. She counts her 2 teenagers as her greatest achievement. Reach her at malisha@ikigaipeople.com.

**Bill Jensen** is a visionary leader, results-driven strategist, problem-solver, globally-known keynote speaker, and author of ten best-selling leadership books, including Simplicity, Disrupt, and Future Strong. As CEO and Chief

Transformation Officer at the Jensen Group, Bill is a trusted advisor to C-suite executives in implementing complex, AI-driven, enterprise-wide changes. His superpower: Translating AI/digital transformation into people transformation. He is globally ranked in the Top Ten thought leaders in the Future of Work and Cultural Transformation. His most valuable valueless possession: a 1950s cardboard box from his childhood home. On the side, in Mom's handwriting: Bill's Mittens, Hats, Scarves. Personal life fantasy: To bicycle around the globe via breweries. Reach him at bill@simplerwork.com.

**Dr. Vivienne Ming** is a neuroscientist, entrepreneur, and author. She's the co-founder and chief scientist of the Possibility Institute, applying machine learning and hybrid intelligence to science and innovation itself. She's also co-founder and executive chair of The Human Trust, an independent nonprofit data trust advancing research in human development while protecting individuals' data. Dr. Ming sits on numerous boards including neurotech startup Optoceutics, UC Berkeley's Neurotech Collider Lab, UC San Diego's Cognitive Science Department, and the Kennedy Family Human Rights Center. She's a visiting professor at UCL's Global Business School for Health and is the author of *Robot-Proof: When Machines Have All The Answers, Build Better People.* Reach her at www.socos.org.

**Anish Padinjaroote** is Group CHRO for Amnex Technologies, and is a human capital and AI transformation leader with 22+ years of global experience across strategy, culture, organizational design, and the future of work. He operates at the fault line between technology and humanity; advising CEOs, CHROs, and boards as AI reshapes how power flows, how work is defined, and how people are valued. He has built and scaled people- and culture-systems in high-growth, AI-led organizations, enabling 5x workforce expansion, while redefining performance, leadership, and employee experience at scale. Anish is the founder of HARE People, a global collective focused on the future of work, and an active voice in global HR and AI leadership communities. Reach him at www.HarePeople.com.

**Nadeesha Wijesingha** works at the intersection of performance, leadership, and inner clarity — helping leaders and organizations unlock results through deeper human insight. Over the past five years, she has held C-suite roles including Chief People Officer and Chief of Staff, partnering closely with CEOs and boards to drive transformation in complex, fast-growing organizations. With over 23 years at Unilever across marketing, sales, and global HR leadership, she brings a rare ability to integrate commercial outcomes with human potential. An architect by training and a Fulbright scholar, she holds dual master's degrees in economics and business. A lifelong practitioner and teacher of mindfulness, Nadeesha believes that sustainable performance is rooted in self-awareness, not just strategy. Her work focuses on helping leaders and organizations grow with both discipline and depth, enabling family businesses to scale like corporates, and corporates to operate with greater purpose.

She is currently writing on leadership and inner transformation, exploring how self-awareness shapes performance and human potential. Find her on LinkedIn @nadeesha-wijesingha, or reach her at charmaneenadeesha@gmail.com.

www.ingramcontent.com/pod-product-compliance
Lightning Source LLC
LaVergne TN
LVHW050303210926
848481LV00014BA/808
*9781972708712*